IMAGES OF WAR

BRITISH TANKS
1945
TO THE PRESENT DAY

A Challenger 2 on the streets of Basra, Iraq. The demeanour of the local on the bicycle suggests that such tanks had become an everyday sight. (*Warehouse Collection*)

IMAGES OF WAR

BRITISH TANKS
1945
TO THE PRESENT DAY

RARE PHOTOGRAPHS FROM
WARTIME ARCHIVES

Pat Ware

Pen & Sword
MILITARY

First published in Great Britain in 2012 by
PEN & SWORD MILITARY
an imprint of
Pen & Sword Books Ltd,
47 Church Street,
Barnsley,
South Yorkshire
S70 2AS

A CIP record for this book is available from the British Library.

ISBN 978 1 84884 566 4

Typeset by Chic Media Ltd.

Printed and bound by CPI Group (UK) Ltd, Croydon, CR0 4YY.

Pen & Sword Books Ltd incorporates the Imprints of
Pen & Sword Aviation, Pen & Sword Family History, Pen & Sword Maritime, Pen & Sword Military, Pen & Sword Discovery, Wharncliffe Local History, Wharncliffe True Crime, Wharncliffe Transport, Pen & Sword Select, Pen & Sword Military Classics, Leo Cooper, The Praetorian Press, Remember When, Seaforth Publishing and Frontline Publishing.

For a complete list of Pen & Sword titles please contact
Pen & Sword Books Limited
47 Church Street, Barnsley, South Yorkshire, S70 2AS, England
E-mail: enquiries@pen-and-sword.co.uk
Website: www.pen-and-sword.co.uk

Contents

Introduction

Despite having designed the first practical tanks during the First World War, the British War Office did little to consolidate its position in this respect during the immediate post-war years and during the 1930s the initiative for tank design passed to Germany. British tanks of the interwar period were generally inferior in most respects to their German counterparts, lacking, particularly, in firepower and protection. Even during the Second World War British tanks were scarcely a match for the German *Panzers* and it wasn't until the appearance of the Centurion in 1945 that Britain was able finally to produce a world-class tank ... and one that, incidentally, might have stood some chance of matching the mighty German *Tiger*.

There was a brief period of hurried mixing and matching of turrets, guns and hulls that led to the appearance of some unsuitable machines, but the immediate post-war period was generally a fertile one for British tank design. The exigencies of the escalating Cold War saw the emergence of two superb machines in the shape of the Centurion and the Chieftain and, while the monstrous Conqueror was rather less than successful, this should perhaps be attributed to an over-reaction to the development of the Soviet IS-3.

Between them, the Chieftain and the Centurion clocked up nearly forty years' service, with the two machines serving alongside one another for six or seven years. But, when the time came to replace the Chieftain, politics once again reared its ugly head and, as in the 1920s and 1930s, defence spending was put on hold. When it finally appeared in 1983, the Challenger 1 was actually little more than a reworked Chieftain with a new turret, the design of which had already been paid for by the Shah of Iran in the shape of the Royal Ordnance Factory's Shir 2. The Ministry of Defence had been planning to replace the Chieftain with the Anglo-German MBT-80 but, when this fell through, it was fortunate that the Shir 2 design was available following its cancellation as a result of the 1979 revolution in Iran. The new design saw its first combat service during the liberation of Kuwait in 1991, before going on to serve in Bosnia, Herzegovina and Kosovo.

Britain's current main battle tank, the Vickers Defence Systems' Challenger 2, is a wholly new design, more than adequately armoured and possessed of considerable firepower. The Challenger 2 should be considered among the best tanks in the world, able to hold its head up high in the exalted company of the US Army's M1 Abrams, the French *Leclerc*, the German *Leopard* and the Israeli *Merkava*. Beyond its name, it owes very little to Challenger 1 and has acquitted itself extremely well in various peace-keeping missions, as well as being used in combat in Iraq, where it provided fire support for the British troops in Basra. To date, despite a request by the current head of the army, General Sir Richard Dannatt, to deploy Challenger 2 in Afghanistan in 2006, the MoD has decided the terrain there is 'not suitable for the Challenger'.

We must also address the vexed question of what actually constitutes a tank. Strangely, there seems to be no official definition beyond the broadest description which states, for example, that a tank is 'a self-propelled heavily armed offensive vehicle having a fully enclosed revolving turret with one major weapon' . . . a definition that would exclude many of the machines which have fought as tanks since the first example appeared, lacking any form of 'revolving turret', in 1916. However, most would agree that a tank is a tracked armoured fighting vehicle (AFV), designed primarily to destroy enemy ground forces by direct fire; such a definition dictates that we must also consider the British Army's CVR(T) and the Warrior MICV as tanks. Finally, it is also customary to describe armoured engineers' vehicles as 'tanks' simply because they tend to be based on tank chassis; for this reason the book also includes armoured recovery vehicles, engineers' assault vehicles, bridgelayers and other similar vehicles.

Chapter One

The Development of the Tank

In September 1915 the first workable prototype for a tracked armoured vehicle was produced by William Foster & Company, a British agricultural engineering company based in Lincoln. Known as 'Little Willie', the machine was quickly followed by a second prototype, dubbed 'Mother' or 'Big Willie'. The latter was sufficiently successful that Fosters and the Metropolitan Amalgamated Railway Carriage & Wagon Company Limited were asked to build 175 units. Although they were officially designated as 'tank, Mk I' for reasons of security, the machines were initially described as 'water carriers for Mesopotamia' – thus giving rise to the name 'tank'.

Looking nothing like the tanks with which we have become familiar, these early machines consisted of a huge rhomboidal box-like hull of riveted boilerplate, with unsprung steel tracks wrapped around the perimeter. Tanks saw their first action at the Battle of Flers-Courcelette on 15 September 1916. The battle lasted for one week and, although expectations were high that the deployment of the tank would prove decisive, the performance of the vehicles was erratic and it would be fair to say that the operation had mixed success. Many felt that the tank needed more improvement before it could be considered ready for use on the battlefield and the design of the machine evolved rapidly following this first deployment.

By 1917 the Mk VIII was being constructed to a standardised design in both Britain and the USA, but it had quickly become obvious that the massive, heavy tanks that had evolved from the Mk I were not suitable for every application. 'Medium' tanks started to appear at the beginning of that same year, against a War Office requirement for a lighter, faster machine that would complement the slower 'heavies'. The first of these, the twin-engined medium Mk A or 'Whippet', armed with the turret from an Austin armoured car, went into action in March 1918, covering the retreat of the British infantry divisions that were recoiling from the German onslaught of the Spring Offensive.

The signing of the Armistice saw tank production all but abandoned in Britain. Although some new tank designs appeared during the interwar period, with the heavy tanks of the First World War being superseded by much lighter, more agile designs, the development process never seemed to be more than half-hearted. By the mid-1930s Germany had probably gained the upper hand in tank design, producing light tanks that were designed to support infantry and by 1939/40 Germany's Blitzkrieg tactics had shown what tanks and infantry could achieve in the hands of well-trained commanders who understood their strengths and weaknesses.

Things had taken a slightly different turn in Britain. In 1938 British War Office doctrine settled on two types of tank, described as 'cruiser' tanks and 'infantry' tanks. Cruiser tanks – sometimes known as 'cavalry' tanks – were seen as medium-weight, fast machines, lightly armoured and lightly gunned, which could make reconnaissance forays deep into enemy territory, much as horse-mounted cavalry had in former conflicts. Infantry tanks, on the other hand, were well protected and lightly armed, usually with no more than a machine gun; the weight of armour meant that they were appreciably slower than the cruisers but, since the role of the infantry tank was to support foot soldiers during an attack, this was considered unimportant. Following this doctrine, new designs, many of which were totally unsatisfactory, were produced during the late 1930s and during the conflict itself. In addition, a shortage of production capacity saw American tanks used by the British Army, including the Stuart M3/M5 light tank, the Lee/Grant M3 medium, and the iconic M4 Sherman medium.

It is interesting to contrast German and British attitudes to tank design during the Second World War. Disregarding captured enemy tanks, between 1939 and 1945 Germany deployed six major types of tank, designated *Panzerkampfwagen I*, or *PzKpfw I* through to *PzKpfw VI*. Modifications were frequently made during the service life of each type, but the new designs that appeared were the result of improvements to firepower, mobility or protection. For example, *PzKpfw I* was a light tank armed with twin 7.92mm machine guns, while the *PzKpfw II* mounted a 20mm gun. The medium *PzKpfw III* of 1937 formed the primary weapon of the German *Panzer* divisions and remained in production until 1943; on its introduction it was armed with a 37mm gun but this was eventually superseded by a 50mm weapon. The medium-weight *PzKpfw IV* and the heavy/medium *PzKpfw V Panther* were both armed with a 75mm gun, as was the prototype of the *PzKpfw VI Tiger*, whereas the production version of the *PzKpfw VI*, and the later *Königstiger* (King Tiger), were both equipped with the fearsome 88mm gun.

In the early years of the conflict the British approach to tank development appeared a very hit-and-miss affair, with each new design apparently owing little to its predecessors. As regards firepower, it would be fair to say that British tanks were generally less well armed and less well protected than their German counterparts. British cruisers dating from the opening years of the conflict were armed only with a 2-pounder (40mm) anti-tank gun and, while this eventually gave way to the 6-pounder (57mm), and then to the 75mm and 76mm weapons of the American M4 Sherman, for most of the war neither Britain nor the USA managed to produce a tank that could take on the *PzKpfw VI Tiger* or the formidable *Königstiger*.

It was not until the end of the conflict that the British Army finally had access to a tank that could compete with the German *Panzers* on more-or-less equal terms: in fact, there were two. The first of these was the Anglo-American Sherman Firefly – essentially an M4 Sherman medium tank into which had been fitted the British 17-pounder (76.2mm) gun; the other was the Comet, a British cruiser tank armed with a 77mm gun. Had the A41 Centurion, which was originally described as a cruiser tank, been ready in time, it might also have helped to tip the balance.

Following VE-Day large numbers of tanks of all types remained in service with the British Army, including quantities of Shermans and Stuarts; some had been converted to specialised roles, while others were placed into storage. Only the Comet, together with the Cromwell cruiser and the Churchill infantry tank, saw any significant service into the immediate post-war years. The appearance of the Centurion in late 1945, followed by its gradual introduction into service the following year, saw the older tanks gradually being replaced. Ultimately the Centurion went on to prove itself one of the best tank designs of the post-war period, seeing combat with the British Army in Korea as well as with the Israeli Defence Force and the Indian Army.

It was at about the time that the Centurion entered service that the British system of referring to tanks by an 'A' number, introduced in 1926, was discontinued and for the next twenty-five years or so tanks were identified by their 'FV' ('fighting vehicle') number. This was a reference assigned by the Fighting Vehicle Design Department (FVDD) and later by its successor, the Fighting Vehicle Research & Development Establishment (FVRDE). While it may not have been the original intention, the FV numbers, which were actually drawing office references, came to be regarded as definitive codes identifying both the families of vehicles and the specific variants. The practice of assigning names to the tanks, which dated back to 1940, was continued and the curious custom of selecting names beginning with the letter C also persisted.

The Centurion was originally assigned the reference A41, but later became known as the FV4000 series. Although it was an excellent design, it was felt that a heavier tank was also required and by 1950 work was well advanced on the colossal Conqueror, a heavy gun tank with a maximum thickness of armour of 178mm, armed with a 120mm gun. Produced in response to the appearance in 1945 of the Soviet IS-3 (a huge machine with a maximum thickness of armour of 132mm, equipped with a hard-hitting 122mm gun), the Conqueror was intended to provide long-range fire and anti-tank support to Centurion regiments. However, it was a victim of its own size and weight, and was never considered satisfactory; eventually the Chieftain, which followed many of the successful design principles of the Centurion, replaced both the Conqueror and the Centurion itself.

By the early 1980s the Chieftain had been superseded by the Challenger 1, the vehicle by now described as a main battle tank (MBT). This, in turn, was replaced by the all-new Challenger 2 after the First Gulf War.

The concept of light tanks had fallen into disrepute before the end of the Second World War, only to be revived in the early 1960s when the aluminium-armoured 'combat vehicle, reconnaissance, tracked' – or CVR(T) – appeared. Designed by the Military Vehicles Engineering Establishment (MVEE) and subsequently licensed to Alvis Vehicles, the basic chassis has been adapted to a number of roles, with the Scorpion, Scimitar and Stormer all equipped for use in what is frequently described as the 'light tank' role. By 1996 more than 3,500 examples had been built for British Army use as well as for export, and many remain in service to this day.

Finally, the GKN Warrior mechanised infantry combat vehicle (MICV) provides a modern application of the infantry tank concept. Originally described as the MCV-80, and resembling a medium-sized gun tank, the vehicle effectively combines two roles, acting initially as an armoured personnel carrier (APC) and then providing supporting firepower from a turret-mounted 30mm RARDEN cannon once the troops have disembarked.

From its first appearance on the Western Front, it had taken almost twenty-five years for the tank to come of age, but there are those who predict that the age of the tank is over and that the machine has become something of an anachronism in the face of today's asymmetric warfare. However, there have been enormous technological advances in various areas which combine to make the modern tank a formidable and well-protected force on the battlefield. NBC (nuclear, biological, chemical) filtration systems have made it possible for crews to survive the deployment of tactical nuclear and chemical or biological weapons. The widespread use of electronics has improved night-vision, target acquisition and fire-control issues, while the development of high-performance anti-armour ordnance (including kinetic-energy long-rod penetrator and depleted uranium penetrator rounds) has made enemy armour considerably more vulnerable. At the same time the use of advanced armour systems such as reactive armour and Chobham ceramic composites has reduced the chances of the modern battle tank being compromised by all but an enemy with comparable technological advancement.

British Challenger tanks were used in both Gulf Wars and, despite reports that the numbers of tanks are likely to be reduced by 40 per cent under the 2011 Strategic Defence and Security Review (SDSR), nevertheless the British Army will continue to have access to more than 200 Challenger main battle tanks for the foreseeable future. Elsewhere, Canadian, Danish and Dutch troops have all used tanks in Afghanistan during the present conflict, and a recent report (November 2010) suggested that the US Army was planning to deploy fourteen Abrams M1A2 main battle tanks to southern Afghanistan, where fighting against Taliban militants is at its fiercest.

Perhaps reports of the death of the main battle tank have been greatly exaggerated!

Dating from September 1915 and built by William Foster of Lincoln, the first workable prototype for a tracked armoured vehicle was dubbed 'Little Willie'. This iconic machine has survived, and forms part of the Tank Museum collection. (*Andrew Skudder*)

Tanks saw their first action at Flers-Courcelette on 15 September 1916. Although the operation was not entirely successful, the Army Chiefs of Staff were sufficiently impressed to order more machines, and development continued apace. In March 1917 five Mk II tanks were used as development 'mules' to try out various transmission systems, with the engine of the tank shown here driving the tracks through a Williams-Janney hydraulic system, using pumps with adjustable swash plates to alter the speed. (*Warehouse Collection*)

The British may have invented the tank but, in the form of the Renault FT-17, it was the French who produced the first tank to mount a revolving turret carrying a heavy gun. Some 1,600 of these machines remained in service with the French Army until at least 1939. (*Warehouse Collection*)

This light tank, the Mk VIA, is typical of the British armour that was in service when the British Expeditionary Force set out for France in September 1939; with thin armour and nothing more lethal than a .50in calibre machine gun, the design was vulnerable to enemy fire and production ended in 1940. (*Warehouse Collection*)

Alongside the quickly discredited light tanks, the British Army also fielded so-called infantry and cruiser tanks, each with a differing role. A small number of examples of the infantry tank, the Mk I or Matilda, also went to France in 1940, but most were confined to training. (*Warehouse Collection*)

Introduced in 1942 and initially armed with a 6-pounder (57mm) main gun, the Cromwell cruiser tank was one of the better British designs of the period, and some remained in service into the post-war years. (*Warehouse Collection*)

The German *Tiger* (*PzKpfw VI*) was a formidable machine. With its 88mm gun and a maximum 100mm of armour, it was better armed and better protected than almost anything the Allies could throw against it. (*Warehouse Collection*)

Introduced in 1944, the *Königstiger* (*King Tiger* or *Tiger II*), was the best protected and most heavily armed tank of the Second World War, but it was both unreliable and not available in large enough numbers. This example has the Henschel turret, which was not adopted for production. (*Warehouse Collection*)

Also dating from 1944, the British Comet was effectively a Cromwell on which was mounted a larger cast turret and a new turret ring, designed to accommodate a new Vickers 77mm gun. An excellent hard-hitting design, the Comet remained in service until 1958. (*Warehouse Collection*)

The Soviet IS-3 heavy tank appeared in 1945 and, with its maximum 230mm thickness of armour and 122mm gun, was a real game-changer. It forced both Britain and the USA at least partially to rethink their approach to tank design. (*Warehouse Collection*)

Armed initially with a 17-pounder (76.2mm) gun, the British Centurion came just too late to have any effect on the outcome of the war in Europe, but it was an excellent design that remained in production until 1962. (*Warehouse Collection*)

Entering service from 1967, the Chieftain replaced both the Centurion and the unsuccessful Conqueror heavy tank, the latter produced in response to the threat of the Soviet IS-3. (*Tank Museum*)

By the early 1980s the Chieftain was being superseded by the Challenger 1. Originally developed for Iran under the name Shir, it was effectively an upgraded Chieftain, powered by a Rolls-Royce V12 engine. (*Warehouse Collection*)

With a 30mm RARDEN cannon, the Warrior infantry fighting vehicle (IFV) is not a tank in the conventional sense of the word. But the vehicle is more than capable of providing fire support to infantry and, with an optional 105mm gun, can also engage enemy light armour. (*Warehouse Collection*)

Chapter Two

Post-War British Gun Tanks

Few could argue that the majority of British tank designs of the Second World War did not compare well to their German equivalents, nor even, in many respects, to the American Stuarts and Shermans. British tanks of the period tended to be under-gunned and inadequately protected, and were often unreliable. The end of the war gave British tank designers the opportunity not only to take stock of the good and bad points of domestic tank production, but also, through the work of the Combined Intelligence Objectives Sub-Committee, to access all German technical research into aspects of tank design and technology. The emergence of the huge Soviet IS-3 main battle tank in 1945 had also caused the Allies to reassess some of the conventional wisdoms regarding levels of firepower and protection.

As might be expected, this led to some revisions in the Allied approach and, although there were some unfortunate cul-de-sacs along the way, notably the Conqueror heavy gun tank, the post-war years saw Britain produce two of the world's best tanks in the shape of the Centurion and the Chieftain. When these became due for replacement there was talk of Anglo-German cooperation in tank design, but when the MBT-80 fell by the wayside it was replaced by Challenger – later to be known as Challenger 1. In truth this was little more than an update of the Chieftain and was superseded in turn by Challenger 2, which provided huge improvements in firepower and fire control. Although the name suggests that Challenger 2 was a development of the previous model, in fact it was an all-new design and was the first British tank to be designed, developed and constructed by the prime contractor, Vickers Defence Systems, with other work carried out under subcontract by the Royal Ordnance Factories.

The 1960s also saw the re-emergence of the light tank in the form of the Alvis combat vehicle reconnaissance tracked, or CVR(T). The light tank role had fallen from favour during the Second World War, with wheeled armoured cars increasingly used for reconnaissance duties. The CVR(T) was an air-portable aluminium-hulled tracked vehicle that was offered in a range of variants, including versions armed with a turret-mounted 76mm or 90mm rifled gun, or a 30mm RARDEN cannon.

Finally, the development of the so-called infantry fighting vehicle (IFV) – or mechanised combat vehicle – showed that it was possible to transport infantry into battle and at the same time to provide fire support from a turret-mounted weapon on the same vehicle. The first purpose-designed machine of this type was the Soviet BMP-1, which appeared in 1967, but since that time all the major military powers have developed, or adopted, IFVs, with the British Army's Warrior a typical case in point. Current thinking suggests that the IFV is more versatile in asymmetric warfare scenarios than is the main battle tank.

Second World War Tanks in Post-war Service

Although work had started on Britain's first real post-war tank in 1944, the first prototypes of what became the A41 Centurion arrived too late to see any action during the conflict and, by May 1945 the British Army's front-line tank force still consisted of wartime infantry and cruiser tank designs, in the form of large numbers of Comets, Cromwells and Churchills, as well as the American Shermans and small numbers of Stuarts that had been adapted to specialised roles. Most of the older British tanks had either been scrapped, passed to Commonwealth units or assigned to training duties, and over the next two or three years there was considerable further rationalisation, with the best of the wartime cruiser and infantry tanks retained, and many surplus vehicles scrapped, converted to other roles or sold to other countries.

However, in May 1946 an official War Office publication listing armoured vehicle nomenclature showed that the fleet still consisted of Cromwell, Comet, Challenger, Churchill, Sherman and Stuart tanks, albeit by this time some may well have been in storage.

In 1948/49 all British military vehicles were renumbered, with the old wartime registrations replaced by a six-digit alpha-numeric system. Remaining Second World War tanks were renumbered, *inter alia*, in the series 00ZR00–99ZR99, and records of the period show that Comets, Cromwells and Churchills still remained in service, along with a handful of Valentines, Sherman beach armoured recovery vehicles (BARVs) and turretless Stuarts, the latter retained for use as gun tractors.

At the same time the A41 Centurion had started to enter service and there was a fresh wave of disposals. In 1948, for example, four Churchill XI gun tanks were 'rented' to the Irish Defence Force for a period of five years, and a number of surplus Comets were offered to various nations, including Finland, Burma, South Africa and Ireland, the latter eventually purchasing a total of eight. Nevertheless, the British Army still retained some Comet gun tanks, as well as the Cromwell close-support variant with its 95mm gun, in their original roles, with some surviving into the mid-

1950s. A number of surplus Cromwells were fitted with a new turret mounting a 20-pounder (84mm) anti-tank gun and were renamed FV4101 Charioteer.

Although the Churchill survived, it was certainly out-gunned by all of the Warsaw Pact tanks and those examples that were retained had already been adapted for specialised roles. These included the Crocodile flame-thrower, bridgelayer, armoured ramp carrier (ARK) and armoured recovery vehicle (AVRE) variants. However, while its gun may have been made obsolete by post-war developments, the Churchill was still relatively well-armoured and for this reason the chassis was adapted for use as a mine-clearing flail during the mid-1950s, with a total of forty-two examples of what was known as the FV3902 Toad built between 1954 and 1956; these are described separately.

In 1941 the British Army started to receive supplies of the US M4 medium tank, better known as the Sherman. The first example to reach these shores, named 'Michael', was displayed in Whitehall and to this day remains in the collection of the Tank Museum. Although flawed and increasingly out-dated, the Sherman was crucial to Allied successes in Europe following D-Day. (*Tank Museum*)

The Sherman was generally considered to be under-gunned when compared to the later German tanks. However, when the British fitted the machine with a 17-pounder (76.2mm) gun, in the form of the Sherman Firefly, it was able to tackle both *Tiger* and *Panther* tanks on a more equal footing, and acquitted itself well in the Normandy campaign. (*Warehouse Collection*)

Although very much a Second World War design, the Rolls-Royce-powered Comet was armed with a big 77mm gun and was fast, agile and hard-hitting. It remained in service well into the 1950s, serving in Korea alongside the heavier Centurion. (*Warehouse Collection*)

The Churchill was another wartime design that saw post-war service, albeit often in specialised roles. Here, a trio of National Servicemen of 19 Company, Royal Army Service Corps, prepare to winch a Churchill off a tank transporter – in 1964! (*Warehouse Collection*)

Many Churchills ended their days as gate guardians. This Crocodile flame-thrower variant stands at the gates of the Muckleburgh Collection in North Norfolk. (*Warehouse Collection*)

Centurion A41: FV4000 Series

By rights, the Centurion should be regarded as a Second World War tank since the first mock-up had been constructed by May 1944 and the first six pre-production vehicles, retrospectively described as the Centurion Mk 1, were delivered in May 1945. But despite being rushed to Germany for a literal baptism of fire, the Centurion arrived too late to participate in the liberation of Europe. Nevertheless, and despite lacking a little in mobility, the type went on to be recognised as one of the best tanks of the post-war period.

The Centurion project had originated with a General Staff Policy Statement on tanks dated 8 September 1943. This emphasised the differing roles of cruiser and infantry tanks, as first laid down in 1938, but also recognised that previous cruiser tanks had been far from satisfactory. There was a call for all-round improvements in the key areas of firepower, mobility and protection, as well as in reliability and durability. At the same time the dimensional restrictions that had forced previous British tank designs to remain within the railway loading gauge were lifted, which provided a considerable degree of design freedom. The main gun was to be a dual-purpose weapon of at least 75mm calibre, capable of firing both high-explosive and armour-piercing projectiles, while the tank was also intended to be capable of accepting larger weapons, for example when adapted for the tank-destroyer and close-support roles.

A month later a more detailed specification was prepared, which stated, among other things, that the proposed future tank should be powered by a 650bhp Rolls-Royce Meteor engine in combination with a Merritt-Brown Z51R five-speed gearbox. It was specified that the maximum weight should not exceed 45 tons, and that the frontal armour should be at least 102mm, with the side armour at least 60 per cent of this figure and preferably more, with the express intention of being able to withstand the German 88mm tank gun. The main gun was to be the existing 17-pounder (76.2mm), together with a co-axial machine gun.

The design work for what was dubbed the A41 was carried out by the Department of Tank Design under Sir Claude Gibb, with AEC eventually appointed as design parent. In early 1944 the Ministry of Supply gave the go-ahead for the construction of twenty prototypes, with a variety of secondary armaments and other detail differences, together with two hulls, one of which was to be of mild steel. The last five vehicles of the series were fitted with a Sinclair-Meadows self-shifting Powerflow transmission in place of the Merritt-Brown unit and were designated A41S. Production was entrusted to the Royal Ordnance Factories at Woolwich and Nottingham, but the soft-skin hull was built at AEC.

The 'soft' hull was ready for inspection at AEC's Southall factory by 24 May 1944 and automotive trials had started by September. Production of the prototypes was authorised to begin in January 1945, with the first vehicle delivered to the Fighting Vehicles Proving Establishment (FVPE) from Woolwich in April.

The increased weight compared to previous cruiser tank designs meant that the Christie suspension had to be replaced by a modified Horstman system, and the maximum speed was held down to little more than 21mph. Unlike previous cruisers, there was no hull gunner's position, and the hull, which was of welded construction, provided space for a crew of four. The 17-pounder (76.2mm) main gun was mounted in a stabiliser (the first time this feature had been included in a British tank), and there was a 20mm Polsten cannon in a co-axial ball mount and a rear-facing 7.92mm BESA machine gun.

Production of the up-armoured A41A Centurion Mk 2 began in November 1945. The thickness of armour was increased to a maximum of 152mm, and the gun was mounted in a new cast turret. These changes increased the weight from 42 tons to 45 tons, and the final-drive ratio was reduced to compensate. The last Centurion Mk 2 was completed in early 1949, by which time Vickers-Armstrong had started work on the Centurion Mk 3, which was armed with a 20-pounder (84mm) gun in a revised turret.

The Centurion entered service with the British Army in December 1946 and saw its first real action in Korea. It remained in British service until as late as 1969, proving itself to be among the best tanks in the world, and was also exported to Australia, Canada, Denmark, Egypt, India, Iraq, Israel, Jordan, Kuwait, the Netherlands, New Zealand, South Africa, Sweden and Switzerland. In all, the Centurion was developed by gradual improvement through thirteen 'marks' and a total of twenty-five variants, with later versions having the 17-pounder (76.2mm) gun replaced by a 20-pounder

(84mm) and then by the 105mm. Production eventually took place at the Royal Ordnance Factories at Woolwich, Leeds and Nottingham, at Leyland Motors and at Vickers-Armstrong, Newcastle-upon-Tyne:

- Centurion Mk 1: armed with a 17-pounder (76.2mm) main gun and a co-axial 20mm Polsten cannon in a partially cast turret; rear-facing 7.92mm BESA machine gun
- Centurion Mk 2: armed with a 17-pounder (76.2mm) main gun and a co-axial machine gun in a cast turret; no rear-facing machine gun
- Centurion Mk 3: armed with a 20-pounder (84mm) main gun and a co-axial machine gun
- Centurion Mk 4: close-support variant armed with a 95mm howitzer; no series production
- Centurion Mk 5: redesigned turret; armed with a 17-pounder (76.2mm) main gun and a co-axial .30in Browning machine gun; designated FV4001
- Centurion Mk 5/1: up-armoured Centurion Mk 5; FV4011
- Centurion Mk 5/2: as Centurion Mk 5, retro-fitted with 105mm L7 gun
- Centurion Mk 6: up-armoured Centurion Mk 5 with long-range fuel tanks and 105mm L7 gun
- Centurion Mk 6/1: as Centurion Mk 6, but with infrared night-vision equipment
- Centurion Mk 6/2: as Centurion Mk 6, but with ranging machine gun
- Centurion Mk 7: redesigned by Leyland Motors; armed with a 20-pounder (84mm) main gun; FV4007
- Centurion Mk 7/1: up-armoured Centurion Mk 7; FV4012
- Centurion Mk 7/2: as Centurion Mk 7, but armed with 105mm L7 gun
- Centurion Mk 8: improved version of Centurion Mk 7 with new gun mantlet and independent commander's cupola; FV4014
- Centurion Mk 8/1: up-armoured Centurion Mk 8
- Centurion Mk 8/2: as Centurion Mk 8/1, but armed with 105mm L7 gun
- Centurion Mk 9: up-armoured Centurion Mk 7, armed with 105mm L7 gun; FV4015
- Centurion Mk 9/1: as Centurion Mk 9, but with infrared night-vision equipment
- Centurion Mk 9/2: as Centurion Mk 9, but with ranging machine gun
- Centurion Mk 10: up-armoured Centurion Mk 8, armed with 105mm L7 gun; FV4017
- Centurion Mk 10/1: as Centurion Mk 10, but with infrared night-vision equipment
- Centurion Mk 10/2: as Centurion Mk 10, but with ranging machine gun
- Centurion Mk 11: as Centurion Mk 6, but with ranging machine gun and infrared night-vision equipment

- Centurion Mk 12: as Centurion Mk 9, but with ranging machine gun and infrared night-vision equipment
- Centurion Mk 13: as Centurion Mk 10, but with ranging machine gun and infrared night-vision equipment

There were also a number of specialised variants, including armoured recovery vehicle (ARV), beach armoured recovery vehicle (BARV), 'dozer, canal defence light (CDL), armoured vehicle Royal Engineers (AVRE), bridgelayer and armoured ramp carrier (ARK). The Centurion chassis was also used as the basis for the FV4004 Conway heavy gun tank, the FV4005 self-propelled heavy anti-tank gun, and the FV3802 and FV3805 self-propelled guns. By the time production ended in 1962 some 4,423 examples had been constructed. The Centurion hull was also used as a development vehicle for the Chieftain tank.

Israeli Centurions were upgraded by the use of a Continental AVDS-1790-2A diesel engine, combined with a GM Allison CD-850-6 fully automatic transmission.

The Centurion Mk I first appeared in 1944 in mock-up form, but did not enter production until November 1945. Powered by a Rolls-Royce engine and armed initially with a 17-pounder (76.2mm) gun, it saw its first real action in Korea. (*Warehouse Collection*)

The Centurion was of welded construction and was considerably heavier than earlier cruiser tank designs. In order to support the 42-ton weight, the earlier Christie suspension was replaced by a modified Horstman system. (*Warehouse Collection*)

With an increased thickness of armour that brought the weight up to 45 tons, but retaining the 17-pounder (76.2mm) gun, the Centurion Mk 2 was the first series production version. It remained in production until 1949. (*Warehouse Collection*)

Introduced in 1949, the Centurion Mk 3 was little changed from the earlier Mk 2, but the 17-pounder (76.2mm) gun was replaced by a harder-hitting 20-pounder (84mm) weapon, offering considerable improvement in armour-piercing capabilities. (*Warehouse Collection*)

A Centurion Mk 3 photographed from the rear. The rear-mounted engine was coupled to the rear sprockets via a five-speed Merritt-Brown gearbox. The vehicle is assigned to the Fighting Vehicles Proving Establishment (FVPE) for test. (*Warehouse Collection*)

Photographed in 1960, this is the Centurion Mk 8, an improved version of the redesigned Mk 7, with a new gun mantlet and an independent commander's cupola. (*Warehouse Collection*)

Most National Service tank crews were trained on the Centurion. With the instructor sitting alongside the opened driver's hatch, here we see a Junior Leader driving a Centurion Mk 3 across country during 1965. (*Warehouse Collection*)

With huge rear-mounted air-intake ducts and a snorkel on each exhaust pipe, a Centurion tackles a river crossing. Although the driver's hatch is closed, the periscope remains above the waterline. (*Warehouse Collection*)

A Centurion Mk 3 in Tripoli, in front of the ruins of the Roman amphitheatre at Sabratha. (*Warehouse Collection*)

A forward repair team of the Royal Electrical & Mechanical Engineers (REME), attached to the British Army of the Rhine (BAOR), attends to a disabled Centurion Mk 3. While a Scammell Pioneer SV2S recovery vehicle hoists a replacement Rolls-Royce Meteor above the rear deck, one man struggles with a 'number 19' radio set, and another repairs the damaged track guards. (*Warehouse Collection*)

The Type B barrel of the 20-pounder (84mm) gun had a distinctive fume extractor halfway up the barrel, and was introduced in December 1954. This barrel was eventually retrospectively fitted to most Mk 3 Centurions. This example is privately owned and was photographed at the annual War & Peace Show. (*Simon Thomson*)

A Centurion target tank with a dummy 17-pounder gun. A number of redundant Centurions were converted to this form at Warminster to allow infantry to fire small arms ammunition and dummy anti-tank missiles at them. The tank is heavily armoured to protect the driver. (*Warehouse Collection*)

Charioteer: FV4101

The FV4101 Charioteer tank destroyer was something of a hybrid, consisting basically of a refurbished Second World War late model welded Cromwell VII or VIII hull, to which had been fitted a new lightly armoured powered turret mounting the 20-pounder (84mm) anti-tank gun of the Centurion 3. Alongside the 20-pounder (84mm) gun, there was a co-axial .30 calibre Browning.

The Cromwell had originally been designed by the Birmingham Railway Carriage & Wagon Company, but design parentage was passed to Leyland Motors in late 1941, with the first pilot model delivered in March 1942. Two more pilot machines were completed by the end of the year, followed by a further twenty for training purposes; total production amounted to more than 4,000 units. The Cromwell was one of the better British tank designs of the Second World War, and the model was eventually produced in eight 'marks', with the original 6-pounder (57mm) gun giving way to a 75mm weapon, and with close-support variants mounting a 95mm howitzer. Development of the Cromwell-based Charioteer, which was intended as a tank destroyer, started in 1950 at the Royal Ordnance Factory Woolwich, with the actual conversion work on some 200 production vehicles carried out by Robinson & Kershaw Limited, based in Cheshire. The first example was accepted into service in 1952.

Like the donor Cromwell from which it was derived, the Charioteer was powered by a Rolls-Royce Meteor engine producing something like 600bhp from its 27 litres, installed in conjunction with the Merritt-Brown Z5 combined transmission and steering unit, driving the rear sprockets. The suspension was of the improved Christie type with angled swinging arms suspended on long helical springs, the suspension units being fitted between the twin skins of the hull sides, giving a measure of protection from damage. Five road wheels were fitted on each side, four of which were provided with shock absorbers; there were no track-return rollers. All-up weight was in the order of 28.4 tons and this gave a maximum speed on improved surfaces of 32mph, with 18mph available across country. The hull was 250in long and had a width of 120in.

The vehicles were issued to Royal Armoured Corps units in Germany, although apparently they were only ever used by the Territorials, and were phased out from 1956. Many were subsequently sold to Austria, Finland, Jordan and Lebanon.

The hybrid Charioteer consisted essentially of a Cromwell hull on which was fitted a new turret mounting a 20-pounder (84mm) gun. This rare example is preserved at the Israeli Armoured Corps memorial at Latrun. Charioteers remained in British service until around 1956. (*Bukvoed*)

Conqueror: FV214

The British FV214 Conqueror was a heavy gun tank armed with a 120mm rifled gun, and was designed in direct response to the appearance of the massive Soviet IS-3. It was a logical development of the A45/FV200 universal tank, the design of which had been mooted by Montgomery's 21 Army Group in 1945 as a tank suitable for all situations, and the earliest manifestations of the project were based on an enlarged version of the A41 Centurion chassis, with additional road wheels and suspension units. To this end, the original mild-steel Centurion hull, which had been produced by AEC in 1944, was rebuilt and widened for use as a development vehicle, with the design work undertaken by the newly formed Department of Tank Design (DTD). English Electric was appointed as the main contractor, and the first A45 prototype was completed in 1948.

A year later the A45 project was cancelled, but the chassis was adapted to provide the basis for the FV214 Conqueror. Delays in finalising the specifications for the Conqueror led to the production of the FV221 Caernarvon, a test-bed vehicle which consisted of the hull of the Conqueror on which was mounted the turret of the Centurion 2 (and subsequently the Centurion 3). The Caernarvon allowed experience to be gained in operating vehicles of such extreme size and weight, and it was planned that once sixty Caernarvons had been constructed production would switch to the Conqueror. Both projects proceeded simultaneously despite much changing of minds, and during 1951 Vickers-Armstrong produced two Conqueror prototypes, one in mild steel and one hardened, along with four prototype turrets. The first production Conqueror was finally completed in 1955 and the first twenty vehicles were assigned for troop trials.

Designed for a crew of four, the Conqueror was a huge machine, measuring up at 456in in length, with a total width of 157in. The hull was of all-welded construction, with a maximum thickness of 178mm, and the turret was a one-piece casting, giving an all-up weight of more than 64.7 tons. Power came from an upgraded version of the Rolls-Royce V12 Meteor, designated M120 No 2 Mk 1A, developing 810bhp at 2,800rpm from 27,022cc by means of a fuel-injection system. Two different transmissions were used, the Z52 and the Z52R; both were of the Merritt-Brown type, and both offered five forward speeds and two reverse – but somewhat unhelpfully the Z52R's gearshift pattern was laid out in a mirror image of the Z52's! With its horizontal coil-springs acting in opposed pairs, the Horstman suspension was similar to that used on the Centurion, but with one extra wheel station and with the benefit of resiliently mounted steel-rimmed road wheels. Unfortunately the decision had been taken to adopt one-piece cast units and these soon exhibited cracks in service, leading to the replacement of every suspension unit that had been constructed. The huge 31in-wide manganese-steel tracks were driven via rear sprockets.

The 120mm L1 gun was derived from the American 4.7in T53 anti-aircraft weapon, with a separate projectile and charge, the latter being brass-cased rather than bagged, allowing space for just thirty-five shells inside the hull. The gun was capable of firing high-explosive (HE), high-explosive squash head (HESH), armour-piercing discarding sabot (APDS) and discarding sabot practice (DS PRAC) rounds and, in theory, it would have been more than capable of penetrating the armour of the Soviet IS-3 had it ever been called upon to do so. There were also two .30in Browning machine guns, one mounted co-axially with the main gun, the other on the commander's cupola. The fire-control system was considered very sophisticated for the period, and the commander's cupola was designed to rotate independently of the turret. This allowed the commander to select a target using the telescopic sight and rangefinder, before instructing the gunner. When the gunner traversed the turret and gun to the preselected lay he would find the target already under his sights; while this was happening, the commander was free to search for the next target.

Conquerors were constructed at the Royal Ordnance Factory Dalmuir, near Glasgow, with additional work carried out at the Royal Ordnance Factory Leeds, and at the Newcastle factories of Vickers-Armstrong; the hull and turret castings were produced by William Beardsmore and English Steel. The total number of vehicles constructed was just 180, or perhaps 185, between 1955 and 1959, including conversions of the Caernarvon. The first twenty vehicles were of Mk 1 configuration, while subsequent vehicles were designated Mk 2, although the differences were minor and included a single driver's periscope and revised frontal armour. Of the various planned variants, only the FV219 and FV222 armoured recovery vehicles entered production.

Nine Conquerors were issued to each of the armoured regiments of the British Army on the Rhine from about 1956–7, generally as three troops each having three tanks. The vehicle proved unreliable in service, with the electrical system and the Mollins automatic ejector gear, which was intended to keep the turret clear of spent brass cases, proving particularly troublesome. The weight was also a problem, holding the maximum speed down to around 21mph, reducing the range to a mere 95 miles and restricting the use of the tank on roads and bridges. All had been withdrawn by 1966, and many ended up as hard targets, where the weak spots of the hull were mercilessly exposed!

F.V. 201.

Developed as a counter to the Soviet IS-3, FV201 was the gun tank variant of the A45 FV200 'universal tank' series and was a logical development of the Centurion. There was no series production, but the design eventually evolved into the Conqueror. (*Warehouse Collection*)

An FV201 photographed from the rear. Note the distinctive Conqueror-type suspension, with horizontal coil springs acting in pairs. The turret is traversed to the rear, and the 20-pounder (84mm) gun is locked in the travelling position. (*Warehouse Collection*)

An overhead view of the FV201 showing the increased width when compared to the Centurion. (*Warehouse Collection*)

The A45 project was cancelled in 1949, but the hull was adapted to form the basis of the Conqueror FV214 heavy gun tank; this example carries weights on the turret, presumably to simulate the weight of the correct gun mantlet. There was considerable doubt regarding the ability of existing transporters to carry the 65-ton machine, but the Thornycroft Antar proved itself to be perfectly capable. (*IWM, 28478/6*)

Two prototype Conquerors were produced in 1951, with production starting in 1955. Following their withdrawal in 1966, many ended their days as very impressive gate guardians. (*Warehouse Collection*)

Even without any reference to scale, this illustration from the military User Handbook admirably shows off the massive size of the Conqueror, all examples of which were produced by the Royal Ordnance Factory Dalmuir. (*Warehouse Collection*)

Although none saw any real action, a Conqueror on the move must have been an awe-inspiring sight, with more than 800bhp of thunderous fuel-injected power driving the tracks. All of the vehicles were based in Germany, and this example, serving with 1st Royal Tank Regiment, was photographed in Hohne. (*Tank Museum*)

Total Conqueror production reached 180, and those examples not fortunate enough to survive as gate guardians after withdrawal made admirable range targets. (*Warehouse Collection*)

(*Above*) Believe it or not, there is more than one Conqueror in private hands. This example was rescued from gate-guardian duties in the 1990s, restored by RR Services in Kent and sold to a collector in the USA. (*Warehouse Collection*)

(*Above and right*) Another privately owned Conqueror recently changed hands through the offices of Army Jeeps, a military vehicle dealer based in Ohio in the USA. It was said to be in good running condition, if a little thirsty! (*www.armyjeeps.net*)

This Conqueror gate guardian can still be seen outside the Stanley Barracks of the Royal Tank Regiment at Bovington in Dorset ... and it looks every bit as impressive in the metal! (*Warehouse Collection*)

Chieftain Main Battle Tank: FV4201

During the early 1950s the British Army planned to replace both the Centurion and the unsuccessful Conqueror tanks with a new design. In 1956 Leyland Motors, which had been appointed the design parent for the Centurion Mk 7, constructed three prototypes of a new tank, which was designated FV4202. The design was similar to the Centurion but the height of the hull was reduced (by placing the driver in a reclined position) and there was a new turret, lacking a mantlet. In 1958 the War Office issued a statement of characteristics for a new British tank that drew heavily on Leyland's proposals and work began later that year.

Most of the design work was carried out by Leyland, but work on the turret was entrusted to Vickers-Armstrong. The first mock-up for what would become known as the FV4201 Chieftain appeared in 1959 and small changes were made to the specification; the first prototype was ready later that year. A further six prototypes were constructed between July 1961 and April 1962; following extensive trials, the Chieftain was accepted for service in May 1963, with production lines established at Vickers-Armstrong's Elswick plant and at the Royal Ordnance Factory Leeds. The first examples did not enter service until 1967, but the Chieftain served as the main battle tank of the British Army for more than two decades, with the last examples leaving service in 1995.

The familiar Rolls-Royce Meteor engine of the Centurion and the Conqueror was eschewed in favour of a Leyland L60 unit, a vertically opposed six-cylinder two-stroke multi-fuel diesel engine producing 685bhp, and later upgraded to give 700–720bhp, from 11,365cc. The engine was coupled to a Merritt-Wilson, later David Brown, TN12 six-speed semi-automatic transmission and steering unit, driving the rear sprockets.

With a combat weight of 52.7–54 tons, the hull of the Chieftain was among the best protected of the period, consisting of cast sections welded together; the adoption of a single casting at the nose eliminated the weak spot on the Centurion where the driver's roof joined the glacis plate. Overall, the hull measured 296in in length, with a width of 138in; the height was some four or five inches less than the Centurion. The turret consisted of a well-shaped front casting combined with rolled plate at the rear; the conventional mantlet was dispensed with, both to reduce the overall weight and improve the protection. Chieftains were subsequently retro-fitted with Stillbrew *appliqué* steel-ceramic composite armour to reduce vulnerability around the frontal arc. The fighting compartment included an NBC (nuclear, biological, chemical) filtration system and automatic fire-detection, and there was fire-extinguishing equipment for the engine compartment.

The main gun was a 120mm L11A5, using a separate bagged charge that allowed it to fire kinetic energy (KE) or chemical energy (CE) rounds, including armour-piercing discarding sabot (APDS), high-explosive squash-head (HESH), smoke, illuminating, and armour-piercing fin-stabilised discarding sabot (APFSDS) rounds. The gun-control system supplied by Marconi provided four modes of control: power, stabilised power, emergency battery and manual. Marconi also supplied the fire-control system (FCS), which included a cupola-mounted periscopic sight for the commander, a pair of sights for the gunner and a fire controller.

The last Chieftains were constructed for Kuwait in 1975. During its fifteen-year production life, the tank was developed through a number of major 'marks', as follows:

- Chieftain Mk 1: original pre-production vehicle, powered by a 585bhp engine; armed with a 120mm L11A5 main gun, 0.5in co-axial machine gun and 7.62mm general-purpose machine gun (GPMG)
- Chieftain Mk 1/2: as Chieftain Mk 1, but with improved cupola and additional 7.62mm roof-mounted GPMG
- Chieftain Mk 1/3: as Chieftain Mk 1, but with 650bhp engine
- Chieftain Mk 1/4: as Chieftain Mk 1, but with 650bhp engine and modified 0.5in ranging machine gun
- Chieftain Mk 2: improved turret and 650bhp engine; this was the first Chieftain to enter service.
- Chieftain Mk 3: improved auxiliary generator, modified commander's cupola with 7.62mm machine gun, and other technical improvements
- Chieftain Mk 3/G: prototype with engine air intake in the turret
- Chieftain Mk 3/2: modified Chieftain Mk 3/G
- Chieftain Mk 3S: production version of Chieftain Mk 3/G
- Chieftain Mk 3/3: Chieftain Mk 3, with modified 0.5in ranging machine gun, Barr & Stroud laser rangefinder, 720bhp engine, improved NBC filtration, and other technical improvements
- Chieftain Mk 3/3P: Chieftain Mk 3 for Iran
- Chieftain Mk 4: increased fuel capacity and minor modifications; only two constructed
- Chieftain Mk 5: as Chieftain Mk 3/3, but with strengthened transmission, modified gunner's telescope and commander's collimator, new exhaust system, new NBC pack, improved ammunition stowage, and other technical improvements. This was the definitive version of the tank
- Chieftain Mk 5/2K: Chieftain Mk 5 for Kuwait
- Chieftain Mk 5/5P: Chieftain Mk 5 for Iran
- Chieftain Mk 6: as Chieftain Mk 2, but with 720bhp engine
- Chieftain Mk 7: as Chieftain Mk 3 or 3S, but with 720bhp engine
- Chieftain Mk 8: as Chieftain Mk 3/3, but with 720bhp engine and modified 0.5in ranging machine gun
- Chieftain Mk 9: as Chieftain Mk 6, but with improved fire-control system

- Chieftain Mk 10: as Chieftain Mk 7, but with improved fire-control system
- Chieftain Mk 11: as Chieftain Mk 8, but with improved fire-control system
- Chieftain Mk 12: as Chieftain Mk 5, but with improved fire-control system

The Chieftain was also developed into an armoured recovery vehicle (ARV) and vehicle-launched bridge (AVLB) for the British Army, and provided the basis for the Rolls-Royce-powered Shir 1 and 2 and the Chieftain 900 main battle tanks, all of which were intended for export.

Chieftain prototype number four (P4) photographed during wading trials at the Fighting Vehicles Research & Development Establishment (FVRDE). One of six prototypes, this particular vehicle was delivered for trials in July 1961 and was the first Chieftain to be used for firing trials. (*Warehouse Collection*)

A Chieftain Mk 5 undergoing desert trials at the Yuma Proving Ground, Arizona. Among other features, the Mk 5 featured a strengthened transmission, a new exhaust system and improved ammunition stowage, and is generally considered to be the definitive version of the design. (*Warehouse Collection*)

Another Chieftain Mk 5, this time undergoing some kind of trials at the FVRDE Chertsey test track. The clue is the white-coated boffin! (*Warehouse Collection*)

A Chieftain being loaded on to the 65-tonne Crane Fruehauf semi-trailer of the Scammell Commander tank transporter. The tank is loaded forwards with the turret reversed to the rear. The Commander replaced the older Thornycroft Antar from late 1984 to become the British Army's only tank-transporter outfit. (*Warehouse Collection*)

The 120mm L11A5 gun fitted to all Chieftains (the photograph shows another Mk 5) was fully stabilised to allow accurate firing on the move, and was carried in a fully traversing turret in a mount that provided an elevation arc of −10 degrees and +20 degrees. The barrel was fitted with a thermal sleeve and a fume extractor. (*Warehouse Collection*)

A Chieftain Mk 2, with an improved turret, of the Blues and Royals fording a shallow river during operations in Germany. The photograph was taken in 1969, soon after the formation of the Blues and Royals from an amalgamation of the Royal Horse Guards (Blues) and The Royal Dragoons (1st Dragoons). (*Warehouse Collection*)

The Chieftain was designed to permit wading to a depth of 42in (3ft 6in) without preparation. Following waterproofing treatment, and with the addition of a snorkel, the depth could be increased to 180in (15ft), which puts the tank completely under water. (*Warehouse Collection*)

A pair of Chieftains of 4th Royal Tank Regiment (4 RTR) photographed at Münster, Germany, in July 1975. 4 RTR merged with 7 RTR in 1959 and saw service in both the Middle East and the Far East. In 1969 the regiment returned to Hohne in Germany where it was issued with Chieftains. (*Warehouse Collection*)

Withdrawn Chieftains made splendid range targets. (*Simon Thomson*)

The Chieftain Mk 11 was an upgrade of the Mk 10 with the turret searchlight replaced by the thermal observation and gunnery system (TOGS). Manufactured by Barr & Stroud, originally as a private venture, the system was adopted in 1979/80 and more than 320 Chieftains were converted to carry the equipment. (*Simon Thomson*)

The Chieftain remained in production for fifteen years. Most went to the British Army, but more than 700 examples were sold to Iran, as well as to Kuwait, Oman and Jordan. The last British Army Chieftains left service in 1995. Many were scrapped or passed to museums and private collectors, but some ended up as hard targets on various firing ranges. (*Simon Thomson*)

As you might expect, the Tank Museum has more than one Chieftain in its collection, including one of the prototypes (99SP23), built in 1959. The example shown here, another Tank Museum exhibit, is a Mk 2 constructed at ROF Leeds and disposed of in August 1997. It is pictured at Tankfest. (*Simon Thomson*)

CHALLENGER 1 MAIN BATTLE TANK

In the late 1950s, with the Centurion and Conqueror clearly due for replacement, the War Office planned to replace both types with the FV4201 Chieftain, which entered production in May 1963. Development of the Chieftain continued through the life of the vehicle and in 1974 Iran placed an order for 125 improved Chieftain Mk 5/5 tanks, designated Shir 1 (the name derived from the Shah of Iran) or FV4032 (sometimes FV4030/2), and 1,225 examples of the improved Shir 2 (FV4033 or FV4030/3). Both types were to be fitted with a Rolls-Royce – later Perkins – Condor CV12 engine and had improvements made to the suspension, sighting system and transmission of the basic Chieftain. However, although production of Shir 1 was well under way and the development of Shir 2 was almost complete, the contract for both models was cancelled in early 1979 following the Islamic revolution and the overthrow of the Shah in Iran. However, the work was not entirely wasted since later that year the Royal Ordnance Factory Leeds received a contract from Jordan for 274 examples of the Khalid main battle tank, which was essentially a development of Shir 1.

Meanwhile, the British Army had reached the point where its Chieftains and any remaining Centurions were themselves due for replacement. Although the original plan had been to adopt the jointly developed Anglo-German main battle tank, this project was cancelled in 1977 and replaced by the MBT-80. Unfortunately, the MBT-80 project was no more successful and the development phase eventually became so extended that there seemed little prospect of the tanks entering service before the early 1990s. When the MBT-80 project was cancelled in March 1977, the British Army was forced to look elsewhere for a replacement for the Chieftain.

The replacement took the form of the Challenger FV4034 (or FV4030/4), retrospectively dubbed Challenger 1 when development of Challenger 2 started in 1988, notwithstanding the fact that there had already been a British Challenger tank dating from 1942.

Design work for the Challenger was carried out by the Royal Armament Research & Development Establishment (RARDE) during 1981, and production started at the Royal Ordnance Factory Leeds in 1982. The tank was effectively a development of Shir 2, using a version of the hull and running gear of the Chieftain. The major differences in appearance derived from the new ballistically shaped turret and the use of Chobham laminated armour (a composite of ceramics, metals and other materials designed to resist both chemical and kinetic armour-piercing rounds). However, the layout of the tank was absolutely conventional, with the driver seated at the front, the commander and gunner on the right-hand side of the turret and the loader on the left. The turret bustle included an NBC (nuclear, biological, chemical) pack that would allow the tank to continue to operate in the most hostile battlefield conditions. At the rear was a huge Rolls-Royce (later Perkins) Condor CV12 TCA turbocharged diesel engine producing 1,200bhp from 26.11 litres, assembled in an easily removable power-pack, complete with David Brown TN37 four-speed automatic transmission and a Borg-Warner torque converter.

The main armament was the Royal Ordnance Factory 120mm L11A5 rifled tank gun as fitted to the Chieftain; Janes, the defence information specialists, reported that later tanks were to be fitted with the RARDE EXP 32M1 high-pressure gun, but it seems that this never happened. The gun was capable of firing high-explosive squash head (HESH), armour-piercing discarding sabot, tracer (APDS-T), and armour-piercing fin-stabilised discarding sabot, tracer (APFSDS-T) rounds, the latter particularly having better penetration characteristics than the old armour-piercing discarding sabot (APDS) rounds. There were also two 7.62mm machine guns, one mounted co-axially with the main gun and one on the commander's cupola, together with a pair of five-barrel smoke dischargers on the turret. Electrical gun-control equipment was supplied by Marconi, and there was a Barr & Stroud thermal observation and gunnery sight (TOGS).

A new hydro-pneumatic suspension, designed by the Military Vehicles Experimental Establishment (MVEE) and Laser Engineering Development, and manufactured by the Royal Ordnance Factory, gave a better cross-country ride than did the Chieftain. There were six aluminium road wheels, with the drive sprockets at the rear and the return sprocket at the front. Total battle weight was 61 tons, and the top speed on the road was 35mph, with the fuel tanks giving a range of 240 miles. The hull measured 455in in length and had a width of 139in.

The initial order, worth £300 million, called for 243 Challengers, with the first handed over in March 1983. Total production eventually topped 420 vehicles, and the last was delivered in mid-1990, all of them produced at the Leeds factory that had been taken over by Vickers Defence Systems in 1986. RARDE remained the design authority for the tank until June 1986, when responsibility for further development was passed to the Royal Ordnance Weapons & Fighting Vehicles Division. Challenger was demonstrated to various Middle Eastern countries but there were no export orders, and Challenger 1 saw service only with the British Army.

Challenger 1 was also adapted to provide an armoured repair and recovery vehicle (CR ARRV), with the first thirty examples produced by Vickers Defence Systems at Newcastle-upon-Tyne in 1985; total production of the CR ARRV amounted to eighty vehicles, forty-eight of which were built at Leeds and thirty-two at Newcastle.

The British Army Challengers saw service in the First Gulf War, with more than half the total fleet assigned to the action in Kuwait. Various purpose-designed improvement kits were fitted to counter the harsh conditions likely to be experienced in the Middle East and changes were made to the sighting and gunnery systems and to the engine and air-filtration equipment. Additional passive and reactive armour was fitted to counter close-range attack from anti-tank weapons. The average availability never dropped below 85 per cent during the campaign and the use of tanks was hailed as 'the decisive battle winner of the land forces campaign', with the Challengers of the British 1st Armoured Division making 'an important contribution to this success'.

Challenger 1 was replaced by Challenger 2 between June 1998 and April 2002, and surplus British Army Challengers were gifted to Jordan, with the first batch of fourteen handed over in late 1999, to replace ageing Centurions.

FV4211 was an attempt to produce a lightweight, well-protected tank by constructing what was effectively a Chieftain hull in aluminium, incorporating Chobham composite armour. The automotive components were taken from a Chieftain Mk 3/3 and two prototypes were built, followed by nine development vehicles. The design evolved into Shir and then into the Challenger 1. (*Warehouse Collection*)

The first running prototype of the so-called FV4211 'aluminium Chieftain' carried this mock-up 'Windsor' turret, so called because of its resemblance to a certain Berkshire castle. The turret consisted of an empty shell that could be ballasted to represent the weight of the final form of the turret and gun. The vehicle was also described as the mobile test rig (MTR, or Miteron). (*Warehouse Collection*)

Khalid was derived from the abandoned Shir I project and was based on the Chieftain hull and turret, but modified at the rear to accept the Rolls-Royce (later Perkins) Condor CV12 TCA turbocharged diesel engine. In all, 274 vehicles were ordered by Jordan in 1974. (*Warehouse Collection*)

Meanwhile the more advanced Shir 2, also consisting of a modified Chieftain hull with a new power-pack but carrying a new angular turret incorporating Chobham composite armour, subsequently evolved into Challenger 1. (*Warehouse Collection*)

Challenger 1, initially designated FV4034, was designed by the Royal Armament Research & Development Establishment (RARDE), with prototypes constructed at ROF Leeds. The design was based on the modified hull and running gear of the Chieftain that had been developed for Shir 2. Production started in 1982. (*Warehouse Collection*)

Total production of the Challenger 1 amounted to some 420 units; although the vehicle was essentially a product of the Cold War, it saw its first real service in the Middle East. (*Warehouse Collection*)

More than 50 per cent of the British Army's Challenger 1 fleet was deployed to Kuwait and Iraq in 1991. Following some frantic upgrading, which included the fitment of auxiliary fuel tanks, it acquitted itself well and not one was hit by an Iraqi tank gun. (*Warehouse Collection*)

One of the problems encountered when operating any tank at speed in desert or dry and dusty conditions is the large amounts of muck thrown up by the tracks. In Kuwait it was necessary to introduce modifications to prevent damage to the engine and air filtration systems. (*Warehouse Collection*)

A Challenger 1 stands outside Stanley Barracks in Dorset, still in the camouflage it wore during operations in Kuwait and Iraq. (*Warehouse Collection*)

Photographed during Operation Desert Storm, this Challenger I wears the distinctive explosive reactive armour (ERA) on the toe plate, with additional passive armour on the glacis and side plates. (*Warehouse Collection*)

A Challenger I undergoing a power-pack change. The complete engine, transmission and cooling system can be removed as a single unit, and a spare power-pack is carried on the rear deck of the Challenger armoured repair and recovery vehicle (CR ARRV). (*Tank Museum*)

Rear view of Challenger 1 showing the twin auxiliary fuel tanks fitted in Kuwait to extend the vehicle's range. Without these tanks, the maximum range is 280 miles on the road. (*Simon Thomson*)

Most of the Challenger 1 fleet was gifted to Jordan, but a small number have ended up in the hands of museums and there are possibly even a couple in private ownership. (*Simon Thomson*)

CHALLENGER 2 MAIN BATTLE TANK

Design work for a successor to Challenger 1 started at Vickers Defence Systems in the mid-1980s. At the time the British Army had not announced that there was actually any requirement to replace Challenger 1, which continued to be operated alongside an upgraded Chieftain variant. However, in 1988 the Ministry of Defence (MoD) announced that it intended to mount a competition to select a new main battle tank, with the four contenders being the French *Leclerc* from Giat Industries, the German Krauss-Maffei *Leopard 2*, the American General Dynamics Abrams M1A2 and the developing Challenger 2. As part of the competition, Vickers was awarded an £80 million contract in 1988 for the construction of nine prototypes, together with two additional turrets. Following a period of evaluation, which ended in September 1990, the MoD announced that Challenger 2 was felt to offer the best combination of performance and through-life costing.

Following a brief interlude, during which Saddam Hussein was forcibly ejected from Kuwait, a £520 million contract was awarded to Vickers in June 1991 for the construction of an initial 127 tanks and thirteen driver-training vehicles; a further 259 tanks were ordered in 1994. Construction took place at a new factory in Leeds, and deliveries were scheduled to begin in June 1994.

Challenger 2 is a completely new design, although the low, angular hull and turret bear a superficial resemblance to Challenger 1. As originally built, the hull was protected by Dorchester armour (a variation of Chobham armour), but it has subsequently been upgraded by the use of add-on explosive reactive armour (ERA) panels manufactured by Rafael Advanced Defense Systems of Israel. Like Challenger 1, the tank is powered by the Perkins (now owned by Caterpillar) Condor CV12 TCA turbocharged diesel engine, in conjunction with a David Brown automatic transmission, in the form of the six-speed TN54 unit. The tank measures up at 452in long and has a width of 138in; its combat weight is 62.5 tons or more. It has a top speed on improved roads of 40mph, with 25mph achievable across country; the range is 280 miles on roads and 150 miles across country. The use of hydrogas suspension provides a stable firing platform regardless of the terrain, and both the turret- and gun-drive systems are electric, thus avoiding the dangers associated with having high-pressure hydraulics in the fighting compartment. The hull also incorporates full heating, air-conditioning and NBC filtration systems.

The main armament is a 120mm L30A1 high-pressure rifled gun mounted in a new turret and capable of firing high-explosive squash head (HESH), armour-piercing discarding sabot, tracer (APDS-T) and armour-piercing fin-stabilised discarding sabot, tracer (APFSDS-T) rounds. Electrical gun-control and stabilisation equipment is supplied by BAE Systems, and the twin 32-bit digital fire-control computer comes from the Canadian company Computing Devices and uses a standard MIL STD1553B databus. There is also a co-axial 7.62mm Boeing chain gun, and a 7.62mm general purpose machine gun (GPMG), the latter fitted to the turret for anti-aircraft protection. Smoke-grenade launchers are fitted to the turret sides, and Challenger 2 can also create a smokescreen by injecting cold diesel fuel into the exhaust system.

In January 2004 the so-called CLIP project (Challenger lethality improvement programme) saw work begin on developing a new smooth-bore 120mm gun for Challenger 2. A single tank was fitted with the Rheinmetall L55 gun, as fitted to the German *Leopard* 2A6 tank, using the cradle, thermal sleeve, bore evacuator and muzzle reference system of the original L30A1 weapon. Firing trials started in 2006 and it has been suggested that it will cost £386 million to fit all British Army Challengers with the Rheinmetall gun.

In addition to the British Army Challengers, a total of thirty-eight vehicles have been supplied to the Royal Oman Army. The hull of the Challenger 2 has also been used to develop the Titan bridgelayer and the Trojan armoured engineer tank (formerly designated AVRE). The Challenger 2E variant, introduced in 2000 for export markets, was powered by an MTU 883 turbocharged diesel engine coupled to a Renk HSWL 295TN five-speed automatic transmission and was controlled by a steering wheel rather than the more conventional tillers. The engine is smaller, which allows the use of larger fuel tanks, increasing the range to a maximum of 340 miles. Marketing of this variant was discontinued in 2005.

From 2002 the production of Challenger 2 was undertaken by Alvis Vickers when Alvis took over Vickers Defence Systems, and then by BAE Systems from 2004 when Alvis itself was taken over.

With the addition of sand filters and various upgrades, Challenger 2 was used during the invasion of Iraq in March 2003, and was particularly important during the siege of Basra, as well as in various peace-keeping missions and exercises. However, the British coalition government's Strategic Defence and Security Review (SDSR) may well see the number of Challengers in service reduced by around 40 per cent.

Despite a superficial resemblance to its predecessor, Challenger 2 was a wholly new vehicle. Designed by Vickers Defence Systems, it is the first British tank to be designed and manufactured exclusively by a prime contractor. (*Vickers Defence Systems*)

A Challenger 2 in Iraq, with additional ERA armour attached to the hull and to the side panels. (*Warehouse Collection*)

A Challenger 2 photographed during manoeuvres on Salisbury Plain. The imaging camera for the TOGS system can be seen on the turret, above the gun barrel with which it is designed to move. (*Andrew Skudder*)

The Challenger 2 played a part in the 2003 siege of Basra. Note the thermal 'friend or foe' combat identification panels (CIP) fitted on the turret sides and rear, developed during the first Gulf War to reduce 'friendly fire' incidents. (*Warehouse Collection*)

Photographed in Basra during Operation Telic, this Challenger 2 has explosive reactive armour (ERA) panels fitted to the toe plate. (*Warehouse Collection*)

There are occasions when nothing beats covering a tank with a scrim net to conceal it from aerial observation. When not required, the netting is rolled up at the front of the hull. (*Warehouse Collection*)

With the benefit of the advanced Dorchester (formerly Chobham) composite armour, plus the addition of both explosive reactive armour (ERA) and slat armour protective guards around the rear of the hull and turret, this Challenger 2 of the Royal Scots Dragoon Guards tries to live up to its description of being among the world's best-protected tanks. (*Warehouse Collection*)

A pair of Challenger 2s photographed in Basra, with CIP identification panels, ERA side armour, dust skirts and rolled-up scrim netting. (*Warehouse Collection*)

ALVIS CVR(T): FV100 SERIES

Although, as the name might suggest, the Alvis 'combat vehicle reconnaissance (tracked)' or CVR(T) is probably more correctly described as a reconnaissance vehicle, it has proved itself to be an extremely versatile machine, capable of adaptation to cover a number of different tasks. Alongside vehicles suitable for the anti-tank, anti-aircraft, ambulance, recovery and command roles, both the Scorpion and Scimitar variants have been marketed by Alvis as 'light tanks', while the extended Stormer was also available with a two-man turret, mounting either a 76mm or 90mm gun.

Work on what was to become the CVR(T) started in the mid-1950s, when the British Army indicated that it had a requirement for an armoured reconnaissance vehicle to cover a number of differing roles. Following a series of studies, the Fighting Vehicle Research & Development Establishment (FVRDE) concluded that, in fact, two separate vehicles were required to fulfil the range of roles described: one with tracks, which became the CVR(T), the other with wheels, which evolved into the combat vehicle reconnaissance (wheeled), or CVR(W). Three aluminium-armoured test rigs were constructed for the CVR(T), designed to finalise various features planned for the finished vehicle, including the possible use of the Rolls-Royce B60 petrol engine of the Ferret, a hydro-pneumatic suspension system and a scaled-down version of the David Brown TN15 transmission of the Chieftain tank.

In September 1967 Alvis was contracted to build thirty prototypes, seventeen of which were of the Scorpion variant, armed with a 76mm low-velocity gun, the others covering six of the other CVR(T) variants. The first prototype was completed in January 1969, with the remainder delivered by the spring of the following year. By this time the Rolls-Royce engine had given way to a militarised and de-rated version of Jaguar's iconic XK engine, designated J60 and producing 190bhp from a capacity of 4.2 litres. The seven-speed TN15 hot-shift semi-automatic tranmission was retained, but the hydro-pneumatic suspension was replaced by a more conventional torsion-bar set-up, with the 17in-wide steel tracks supported on five rubber-tyred road wheels, the first and last of these being fitted with hydraulic shock absorbers.

As with the test rigs, the vehicle was constructed around a three-man compact aluminium welded hull, measuring 190in in length, with a width of 88in. The driver was placed at the front, alongside the engine and transmission, while the fighting compartment was at the rear; some variants were fitted with NBC filtration equipment. A flotation screen could be erected around the perimeter of the hull to allow amphibious operation, with a maximum speed in the water of 4mph. The thickness of armour was not declared but was said to be resistant to attack from Soviet 14.5mm machine guns over the frontal arc, and 7.62mm armour-piercing rounds elsewhere. The combat weight of the Scorpion variant was 17,761lb and the maximum speed on the road was 50mph.

The CVR(T) was approved for service with the British Army in May 1970, with the first production vehicles delivered in January 1972; it was also accepted for service with the RAF Regiment. The first British contract called for 275 Scorpions and 288 Scimitars, and by the mid-1980s a total of 1,863 CVR(T)s had been constructed for the British Army and the RAF.

The Scorpion variant was equipped with a 76mm L23 gun in a two-man turret, capable of firing high-explosive (HE), high-explosive squash-head (HESH), smoke and canister rounds. There was also a co-axial 7.62mm L43A1 machine gun, and later versions had a second 7.62mm general purpose machine gun (GPMG) on a roof-mounted pintle; in the Scorpion 90, which was trialled for the British Army, the 76mm gun was replaced by a 90mm Mk 3 Cockerill gun. The Scimitar, deliveries of which started in March 1974, had a 30mm L21 RARDEN cannon in place of the 76mm gun, capable of firing armour-piercing (AP), armour-piercing secondary effects (APSE), armour-piercing discarding sabot (APDS), high-explosive (HE) and high-explosive incendiary (HEI) rounds.

The larger and heavier Stormer, the length of which was extended by adding an additional wheel station, was originally designed by Alvis as a private venture in the 1970s for use as an armoured personnel carrier (APC). It was selected by the British Army in 1986 to carry the Starstreak high-velocity missile (HVM) anti-aircraft system and other versions of Stormer are also currently available, including a light tank carrying a 30mm RARDEN cannon or 90mm gun, but no other variants have been specified for the British Army.

Scorpions served during the campaign to regain British control of the Falkland Islands in 1982, while the Scimitar has been deployed on UN and NATO peace-keeping missions in the former Yugoslavia and in Iraq, and is currently deployed in Afghanistan.

In 1988 Alvis was awarded a £32 million contract to carry out a CVR(T) life-extension programme (LEP) covering more than 1,300 vehicles. The programme included replacement of the Jaguar engines with a Cummins BTA 5.9-litre 190bhp diesel, and some up-armouring. At the time of writing (spring 2011) the Scimitar remains in service with the British Army, primarily as a reconnaissance vehicle, with a life expectancy of a further four to five years. Samaritan ambulance, Sultan command vehicle and Sampson recovery vehicle variants also remain in British service, but have not been upgraded. Recent rumours in the defence media have suggested that the Ministry of Defence (MoD) has asked BAE Systems, which took over Alvis in 2004, to investigate the possibility of restarting production of the CVR(T) hull.

The CVR(T) has also been purchased by, among others, the armies of Brunei, Honduras, Iran, Ireland, Jordan, Kuwait, Malaysia, New Zealand, Nigeria, Oman, Philippines, Tanzania, Thailand and the UAE. Various upgrade packages have been developed for export markets, including replacements for the engine and suspension system.

Produced by Alvis in an almost bewildering range of variants, all of which have names beginning with 'S', the CVR(T) – or combat vehicle reconnaissance, tracked – started to enter service in 1972. The Scimitar variant (FV107) is armed with the 30mm L21 RARDEN cannon. (*Warehouse Collection*)

The Scorpion variant (FV101) carries a 76mm L23 gun in a two-man turret, and is described as a light tank or a fire-support vehicle. (*Simon Thomson*)

For users requiring additional firepower, the Scorpion could be specified with a long-barrelled 90mm Cockerill Mk 3 gun, in which form it was described as a Scorpion 90 or Scorpion 2. (*Alvis Vehicles*)

The CVR(T) is surprisingly popular as an 'entry level' vehicle for collectors of tracked armour. This 76mm gun-equipped Scorpion was photographed at the War & Peace Show. (*Simon Thomson*)

A Scimitar in full desert-fighting mode, with CIP identification panels and *appliqué* armour panels on the sides of the hull. (*Warehouse Collection*)

An up-armoured Scimitar being put through its paces at the Tank Museum; the 30mm gun is capable of firing a variety of armour-piercing and high-explosive rounds. (*Simon Thomson*)

British Army CVR(T)s have been upgraded by up-armouring the hull and by replacing the original Jaguar engine with a Cummins diesel. This is a refurbished Scorpion variant, with the 76mm L23 gun. (*Simon Thomson*)

Rear view of the Scorpion 90 as trialled for the British Army; there were no series purchases. (*Alvis Vehicles*)

The first British Army regiment to be equipped with the Scorpion was the Blues and Royals in 1973; the last Scorpion was demobbed from the British Army in 1994, but the vehicle remains in service elsewhere. (*Simon Thomson*)

Scimitars remain in British service in the reconnaissance role and may continue to serve until 2015/16. (*Simon Thomson*)

The hull of the Stormer variant was extended by the addition of a sixth wheel station. This example, equipped as an infantry fighting vehicle (IFV), mounts a 30mm gun in an Alvis two-man turret; other weapons available included 20mm, 25mm, 76mm and 90mm guns. (*Alvis Vehicles*)

GKN WARRIOR INFANTRY COMBAT VEHICLE: FV510

Originally described as the MCV-80 (or 'mechanised combat vehicle, 1980'), the GKN Warrior was designed to provide a replacement for the ageing FV430 series of armoured personnel carriers (APCs) – some of which, incidentally, still remain in service at the time of writing (spring 2011) in the form of the much up-graded Bulldog. Strictly speaking, Warrior cannot be considered to be a tank, since it was primarily designed to carry an infantry section into battle. However, by virtue of being available with a variety of turret-mounted weapons, including a 30mm RARDEN L21 cannon, and 75mm, 90mm or 105mm guns, it is also suitable for engaging enemy light armour.

The design work that led to the Warrior was started in the early 1970s when GKN Sankey was awarded a Ministry of Defence (MoD) contract to develop a new infantry fighting vehicle for the British Army. Warrior was selected over the competing American Bradley IFV, with three prototypes delivered in 1980; a production contract was awarded later that year, with the first production machine scheduled for delivery in 1985.

Designed around an aluminium welded hull, Warrior was said to be 'optimised' against close air-burst fragments from 152mm and 155mm weapons and armour-piercing rounds up to 14.5mm; it was also said to have been tested against attack from 9kg anti-tank mines. In its basic 'section vehicle' configuration, it provides accommodation for up to ten soldiers, seven of whom are seated in the rear compartment, with access via double doors at the rear; the remaining three crew members act as commander, gunner and driver. The standard 30mm RARDEN cannon is mounted in a rotating Vickers-manufactured turret, together with a co-axial machine gun. Standard equipment includes an NBC (nuclear, biological, chemical) filtration system.

Power is provided by a Rolls-Royce CV8 TCA V8 diesel engine, latterly described as the Perkins Condor when that company took over the rights to Rolls-Royce diesels. The engine produces 550bhp from a capacity of approximately 17.4 litres, and drives the front sprockets via a Detroit Diesel Allison X300-4B four-speed automatic transmission with two reverse gears; at the time of the vehicle's launch, it was announced that the transmission would be manufactured under licence by Rolls-Royce Limited. Maximum speed on the road is 48mph; a speed of 30mph is available using second reverse gear. Suspension is of the torsion-bar type, with six rubber-tyred aluminium road wheels, produced by GKN's Squeezeform technology that was claimed to provide the strength of a forging with the cost-effectiveness of casting. Hydraulic dampers are fitted to the first, second and sixth wheel stations, and the tracks are of cast-steel with rubber road pads.

In 1983 the defence media publishers Janes reported that the total number of Warriors required by the British Army was between 1,800 and 2,000, although other sources have subsequently suggested that the total number delivered was just 785.

Most were of the basic Warrior FV510 infantry fighting vehicle variant; others include the FV511 infantry command vehicle, the FV512 combat repair vehicle, the FV513 recovery vehicle and the FV514 artillery observation vehicle. A further 254 units of a modified version, known as the Desert Warrior, have been constructed for Kuwait, and the Warrior 2000 was also developed for the Swiss Army but never entered production. Other variants are available, including a low-profile light tank equipped with a 105mm gun, but this version has not served with the British Army.

The manufacturing rights to the Warrior passed to Alvis when that company merged with GKN Defence in 1998, and then to BAE Systems in 2004. Although the vehicle is no longer in production, at the time of writing (spring 2011) BAE Systems continues to provide support and upgrade services.

Warriors have seen action in Iraq and Kuwait during the First Gulf War, and in Bosnia and Afghanistan, where the vehicle has come in for considerable criticism due to the perceived inadequacy of its armour. Nevertheless, the current suggestion is that Warriors may remain in British service until 2025 or even 2035. The British Army recently announced that 643 Warriors were to be upgraded by fitting a modular protection system and enhanced electronics. Around 450 of these vehicles will also be fitted with a new turret and main gun under the Warrior fightability lethality improvement programme (WFLIP); the original RARDEN cannon will be replaced by a stabilised 40mm Anglo-French weapon developed by CTA International.

The GKN Warrior infantry fighting vehicle (IFV) was designed to replace the ageing FV432 APC, and at the same time to be able to provide additional firepower. Although not strictly speaking a tank by most definitions, the availability of a range of turret-mounted weapons means that the Warrior is able to engage enemy light armour. The British Army's FV510 variant mounts a 30mm RARDEN cannon. (*Warehouse Collection*)

Work on the Warrior started in the early 1970s, with the first prototypes delivered in 1980. The 30mm RARDEN cannon was mounted in a Vickers-designed turret. (*Warehouse Collection*)

Photographed in the Middle East during the first Gulf War, this Warrior wears *appliqué* armour panels on the hull sides. (*Warehouse Collection*)

A Warrior travelling across country at some speed. The big Rolls-Royce engine could propel the machine at 48mph on the road. (*GKN Defence*)

A manufacturer's portrait shot of the so-called Warrior desert fighting vehicle (DFV), designed for export sales. (*GKN Defence*)

A veteran of the Gulf War, this Warrior wears appliqué armour on both the nose and the side panels. (*Warehouse Collection*)

Deliveries of the Warrior to the British Army started in 1985, and it is said that the vehicle may well remain in service for another fifteen or twenty years. The type has seen active service in Iraq, Kuwait, Bosnia and Afghanistan. (*Simon Thomson*)

Chapter Three

Experimental Types

Between the mid-1930s and the end of the Second World War, the British Army deployed around twenty types of tank, many of which were notable only for their lack of suitability to the task in hand. Following the end of the war, Churchill's so-called 'iron curtain' descended across Eastern Europe and the Soviet Union was considered to be the new foe. The lessons of the Second World War were well and truly learned, and over the following decades there have been just five types of main battle tank – Centurion, Chieftain, Challenger 1 and 2, and Conqueror – of which only Conqueror was an unqualified failure.

The situation was far less clear following VE-Day and plenty of Second World War types remained in service in those immediate post-war years, but it must have been obvious to all but the most casual observer of the military scene that most of them had less than a snowball's chance in hell against the might of the Soviet IS-3 main battle tank. In the eyes of the West, the 122mm gun of the IS-3, combined with cast-steel armour with a maximum thickness of 230mm, was a real game-changer. The 17-pounder (76.2mm) gun of the Sherman Firefly or the 77mm gun of the Comet might have stood some chance of penetrating the armour of a German *Tiger*, but this Soviet leviathan was another matter altogether.

Inevitably the IS-3 caused something of a panic in the West and suddenly bigger was better. It must have seemed as though everything that had gone before was obsolete ... the FV301 light tank project, for example, became a casualty of this obsession with size even before prototypes had been constructed, and there is no doubt that the appearance of the IS-3 was responsible for the development of the British Conqueror. However, during the six years that it took to get Conqueror into production, the War Office asked the Department of Tank Design whether or not there was any possibility of quickly producing a heavy gun tank that might be capable of taking on the IS-3. The answer should have been a resounding 'no' but several brave attempts were made, often developed along the lines of 'what would happen if we tried the turret of this tank on the hull of that'. However, none entered series production and all were superseded by the Centurion in its various roles.

CAERNARVON MEDIUM GUN TANK: FV221

In April 1950, with the Conqueror heavy gun tank project mired in delays, the Department of Tank Design (DTD) completed the design for an interim medium gun tank, designated FV221 and named Caernarvon. It was effectively the hull of the FV214 Conqueror combined with the turret and gun of the Centurion, mounted via an adaptor ring. With the Conqueror turrets not ready for production, it was felt that the Caernarvon would give drivers an opportunity to get used to handling such a large vehicle.

The first example was prototyped by Vickers-Armstrong in Mk 1 configuration, armed with the 17-pounder (76.2mm) gun of the Centurion 1. A further twenty vehicles were eventually built in Mk 2 configuration, with the 20-pounder (84mm) gun of the Centurion 3, by the Royal Ordnance Factory Leeds, at a total price of £1.4 million. The first of the production vehicles was completed in April 1952 and the tanks were issued for troop trials a year later, and at least one vehicle had the turret ballasted to simulate the weight of the Conqueror turret.

The subsequent success of the Centurion led to the cancellation of the Caernarvon project after the completion of the troop trials in October 1953. Once the Conqueror turrets became available, seven of the Caernarvon hulls were eventually reworked into the standard Conqueror configuration, but one (07BA70), with the turret removed, was fitted with a Parsons gas-turbine engine in 1954, in place of the Rolls-Royce Meteor. The first armoured fighting vehicle (AFV) to be so equipped, it subsequently wound up being used as a dynamometer test vehicle at Christchurch and survives, *sans* gas turbine, at the Tank Museum.

CENTURION 180mm/183mm ASSAULT GUN: FV4005

With work starting in 1951, pending the development of a similar self-propelled gun on the Conqueror chassis, FV4005 was an attempt to mount a 180mm gun on the Centurion Mk 3 hull. Stage 1 of the project consisted of an open-topped hull, with the gun having a limited traverse and a concentric recoil system; in this incarnation the gun was fitted with an auto-loader. In Stage 2 the gun was mounted in a light, splinter-proof turret with a conventional recoil system; loading of the ammunition was by hand.

By December 1952 the original 180mm gun had been replaced by a 183mm weapon but the project did not progress beyond the basic feasibility stage and by August 1957 had been abandoned without any series production. One of the prototypes is possibly still retained at the Royal Military College of Science at Shrivenham.

OTHER CENTURION VARIANTS

In the early 1950s several proposals were made to use the Centurion chassis as a self-propelled (SP) gun mount.

FV3802 was based on a shortened version of the Centurion Mk 7 and was equipped with the venerable QF 25-pounder (87.6mm) gun. The first of three prototypes appeared in October 1955, but the vehicle was not considered satisfactory and, under pressure from the Royal Artillery, became the FV3805, in which role it was equipped with a huge 5.5in gun. Two prototypes were constructed in this form and were trialled, but the project was eventually cancelled in 1960 in favour of the FV433 Abbott.

Although it does not appear that any of these projects progressed beyond the stages of feasibility discussions and/or mock-ups, there were also plans to use the Centurion chassis to mount the 7.2in howitzer (FV3806), the 120mm anti-tank gun (FV3807), the 20-pounder (84mm) medium gun (FV3808) and the 155mm gun (FV3809). All were quickly discounted.

In 1967 British Aerospace demonstrated a Centurion Mk 5 that had been equipped with the Swingfire wire-guided anti-tank missile system, in the form of twin launchers mounted on the turret sides. Both the Centurion and the Chieftain were also used as a mount for the Marconi Marksman anti-aircraft turret.

CONQUEROR GUN TANK VARIANTS

The Conqueror prototype number three was intended to demonstrate the flame-thrower role and consisted of a Centurion Mk 3 turret with a 20-pounder (84mm) main gun and flame-projector equipment. By the time the flame equipment was ready for trials in July 1948, the decision had been taken to abandon the project and to fit the flame-thrower equipment to the Centurion instead.

FV205 was a proposal for mounting a medium anti-tank gun on the Conqueror hull but it was cancelled in April 1949 with little progress having been made. Some consideration was also given to using the Conqueror hull to mount a high-velocity anti-tank gun in a huge ball mount in the glacis plate (FV206), rather in the style of the German tank killers (*Sturmgeschütz*) of the Second World War. This project was abandoned in July 1948, as was a similar project designed to provide a Conqueror-based self-propelled gun using a 152mm weapon (FV207).

In May 1952 there was also an abortive proposal to mount a 120mm medium anti-tank gun on the Conqueror hull under the designation FV217. It had been abandoned by the end of the year.

HEAVY GUN TANK NUMBER 2: FV215B

Development work on what was known as 'heavy gun tank number 2', or FV215B, was carried out by Nuffield Mechanizations & Aero during 1950 with the intention of mounting a 180mm gun on the hull of the Conqueror. Three trial vehicles were intended to be constructed by Vickers-Armstrong, with the work being undertaken between 1951 and 1955, but two of these were subsequently cancelled before the whole project was abandoned in early 1957 to be replaced by the Malkara wire-guided anti-tank missile, mounted on the armoured 1-ton Humber truck chassis.

The same 180mm gun was also mounted in a Centurion chassis under the project designation FV4005.

CONWAY TANK DESTROYER: FV4004

Developed during the period 1950–1952, the Conway FV4004 tank destroyer consisted of the hull of the Centurion on which was mounted a larger gun in an attempt to provide sufficient firepower to counter the Soviet IS-3 heavy tank until work on the Conqueror project was complete.

A single experimental vehicle was constructed by the Royal Ordnance Factory Leeds, carrying a huge rolled-steel turret, designed by the Auster Aircraft Company and constructed by Chubbs of Wolverhampton, in which was mounted the American 120mm L1A1 anti-tank gun intended for the Conqueror. The gun had to be mounted high in the turret to prevent the recoil from impacting on the turret ring, and yet the maximum elevation was just 10 degrees. The height of the turret upset the centre of gravity of the vehicle and made transportation very difficult.

Trials continued throughout 1952, but at the end of the year the Conway project was cancelled. The prototype resides at the Tank Museum.

The Caernarvon (FV221) consisted of the hull of the Conqueror heavy gun tank on which was mounted the turret and gun of the Centurion. A total of twenty examples were constructed, seven of which were subsequently fitted with the Conqueror turret. (*Warehouse Collection*)

In its first incarnation the FV4005 Centurion assault gun consisted of an open-topped Centurion Mk 3 hull on which was mounted a 188mm gun with an auto-loader. (*Warehouse Collection*)

Stage two of the FV4005 project saw the 180mm gun replaced by a 183mm weapon, this time installed in a huge rotating turret. In effect, it was little more than a splinter-proof steel enclosure. This vehicle has survived and is on display at the Tank Museum. (*Warehouse Collection*)

Both the Centurion and the Chieftain (seen here) chassis were used as a mount for the Marconi Marksman anti-aircraft turret. The Marksman system consisted of two 35mm anti-aircraft guns, together with surveillance and tracking radar, carried in a large armoured turret. (*Simon Thomson*)

The development of FV215B was carried out by Nuffield Mechanizations & Aero during 1950 with the intention of mounting a huge 180mm anti-tank gun on the Conqueror hull. The photograph shows a wooden model of the proposed vehicle; a full-scale mock-up was almost completed by mid-1955 before the project was abandoned. (*Warehouse Collection*)

The Conway tank destroyer used the Centurion hull on which was mounted a new turret carrying the 120mm anti-tank gun of the Conqueror. The project was abandoned in 1952, but the prototype survives. (*Warehouse Collection*)

Never making it much beyond the mock-up stage, the Vickers-built FV301 (also known as the A46) was a proposed light (21 ton) tank with a 77mm high-velocity gun. It became a victim of the 'bigger is better' thinking that resulted in the appearance of the Conqueror. (*Warehouse Collection*)

Chapter Four

Engineers' Tanks

So-called 'special tanks' first appeared in 1917 when modified Mk IV heavy tanks were equipped with fascine bundles or hollow timber cylinders to allow ditch crossing. Both Mk IV and Mk V tanks were also fitted with hinged ramps, thus creating the first bridging tanks; others had their armaments removed and were adapted for use as supply vehicles or gun carriers, while the armoured recovery vehicle was developed by the simple expedient of attaching a jib and pulley block or powered crane to the front of an obsolete tank.

The outbreak of the Second World War saw a resurgence of interest in using armoured vehicles for specialised roles, including mine clearance, recovery, demolition, earth-moving or 'dozing, bridgelaying, etc., and, in the lead-up to D-Day a range of so-called 'funnies' was developed, each tasked with overcoming a particular problem. These vehicles made an enormous contribution to the success of the landings and similar machines have remained a vital component of armoured combat units to this day.

Although the range of roles for these special armoured vehicles remains much the same as it was back in 1944, over the last decade or so there has been a tendency to produce multi-purpose machines and the British Army's current range of engineers' tanks includes just three types: the armoured repair and recovery vehicle, the bridgelayer and the earth mover.

CHURCHILL TOAD FLAIL: FV3902

The Churchill infantry tank was originally devised by the Belfast-based Harland & Wolff concern under the designation A20, but most of the real development was carried out by Vauxhall Motors. The first pilot model was completed by November 1940 and the tank went into production as the A22 under a consortium of manufacturers led by Vauxhall. In reworking the design, Vauxhall retained the basic Harland & Wolff hull and the Bedford flat-twelve engine, producing 325–350bhp from a capacity of 21,240cc; drive was made to the rear sprockets via a Merritt-Brown five-speed gearbox and epicyclic steering unit, although after a hundred vehicles had been constructed, this was replaced by a four-speed box. The tracks were wrapped around the perimeter of the hull, and the road wheels were supported on sprung bogie suspension. The hull was of composite construction, consisting of an outer covering of armour plate bolted to a mild-steel inner skin, giving a maximum thickness of armour of 102mm, later increased to 152mm. The first version was fitted with a cast turret armed with a 2-pounder (40mm) gun, with a 3in howitzer mounted in the nose. The howitzer was soon deleted, and the 2-pounder gave way to a 6-pounder (57mm), and then to a 75mm gun, in a larger turret of either cast, welded or composite construction.

The Churchill remained in production until October 1945 and was developed through a total of eleven 'marks'. It was never more than a makeshift design but the spacious and well-protected hull tended to make it a favourite for conversion to specialised roles, including armoured recovery vehicle (ARV); deep-wading beach armoured recovery vehicle (BARV); three types of Ark armoured ramp carrier; hydraulic bridgelayer; bobbin mat layer; various types of mechanical and explosive mine-clearance device; and the Ardeer Aggie and Woodpecker *pétard* mortar projectors. The last Churchill was constructed in October 1945 but, with a total of 5,640 having been built, large numbers inevitably remained in service or in store in the immediate post-war years.

Following VE-Day in May 1945 plans were under way to replace the Churchill and other wartime designs with the Centurion but no provision had been made for a mine-clearance vehicle on the Centurion chassis and, with hundreds of Churchill gun tanks in store, it must have seemed logical to use the Churchill VII hull as the basis for a new flail.

The Distington Engineering Company of Workington was appointed as design parent, with Robinson & Kershaw Limited of Dunkinfield, Cheshire, carrying out the conversion work, which entailed removing the turret and gun, increasing the frontal armour to 140mm and fabricating a new superstructure that would provide an elevated driving position as well as housing a fuel-injected M120 Meteor V12 engine specifically to power the flail drum. A mechanical governor was fitted to maintain the rotor speed at about 150rpm, and the crew was provided with a gyro-stabilised direction finder, together with an electro-magnetic station-keeping device.

The flail assembly consisted of a pair of arms carried in bearings on either side of the hull to support the flail drum, which carried sixty chain flails. Each arm, which was folded back across the tank when not in use, was supported on what looked like a massive hydraulic ram but was actually a spring-balance cylinder. Drive from the flail motor was transferred to the drum via a shaft running inside the left-hand arm, and was passed through the centre of the drum to an epicyclic reduction final drive on the opposite side. The British Rail workshops at Horwich fitted the flail assembly and mechanical components.

At the rear there was a lane-marking system, consisting of an endless belt holding fifty-nine marker poles intended to be fired into the ground at 50ft intervals on one or both sides of the tank during flailing. The mechanism was driven by a power take-off from the final drive, and the poles, which were telescopic, had a blank .303in cartridge in the top that was fired by a hammer built into the mechanism.

A total of forty-two Toads were built between 1954 and 1956, consisting of two prototypes, six pre-production vehicles and thirty-four production models. The vehicle was only ever used for training, however, and only one Toad is thought to have survived.

Squat and ugly like its namesake, the Churchill Toad (FV3902) was an attempt to update the flail concept that had so successfully dealt with German anti-tank mines following the D-Day landings. Although a total of forty-two were constructed, the vehicle was never used in anger. (*Warehouse Collection*)

One Churchill Toad has survived; restored to full working condition, it forms part of the collection of military vehicles of the late Jacques Littlefield. Demonstrating the flailing operation in loose straw, the Toad is seen here before being passed to its new owner. (*Warehouse Collection*)

ARMOURED RECOVERY VEHICLES

Although so-called 'salvage tanks' had been developed during the First World War, the development of the first proper armoured recovery vehicle (ARV) can be dated back to February 1942, when the British Army began to look into the possibility of converting tanks to provide fitters and vehicle engineers with a method of retrieving damaged tanks from the battlefield under some degree of armoured protection. The Royal Electrical & Mechanical Engineers' (REME) workshops at Arborfield produced experimental ARVs based on various tanks, including Covenanter, Crusader, Churchill, Cavalier, Centaur, Grant and Sherman chassis, by the simple expedient of removing the turret and gun and equipping the hull with basic recovery equipment. At one time it was felt that it would improve efficiency if there was an ARV variant of every gun tank, since this would simplify parts supply; production of the Churchill ARV Mk I started in mid-1942.

The following year the British Army had also started to receive examples of the US Army's T2 tank recovery vehicle, based on the M3 Lee hull. This led to the development of a similarly equipped British vehicle, designed by the REME Experimental Recovery Section (ERS) using both Sherman and Churchill hulls. Known as the ARV Mk II, it had a fixed turret mounting a dummy gun and providing space for a Croft 60-ton winch. There was a detachable 3.5-ton winch at the front and a fixed 9.5-ton winch at the rear, as well as a substantial earth anchor. Many of these remained in service into the post-war years.

The British Army also used the Sherman-based M32 tank recovery vehicle, describing it as the ARV Mk III and again some of these saw post-war service.

Centurion armoured recovery vehicle: FV4006

When the 52-ton Centurion Mk 3 started to enter service in about 1950, it was clear that a new recovery vehicle would be required; as an interim measure a number of damaged Centurions were converted to tugs. In this form some were used in Korea. Meanwhile, work had started on designing a purpose-made armoured recovery vehicle using the Centurion hull, with a prototype produced by the REME Command Workshop at Aldershot during 1951. Similar in concept to the Churchill ARV Mk 2, it had a dummy turret and gun, and an 18-ton winch, but the crew compartment was cramped owing to the need to provide a separate winch engine. A total of eleven units were constructed while development work continued on the Mk 2 (FV4006), of which around 170 units were eventually built. The Bedford petrol engine that had been used to drive the 30-ton winch on the Mk 1 variant was abandoned in favour of an electric motor driven by a Rolls-Royce B81-engined generator. Improvements were also made to the layout of the winch and its roping arrangements, and a huge new spade anchor was fitted.

The first Centurion ARV entered service in 1956 and many saw more than thirty years' service.

A proposed Centurion ARV Mk 3 (FV4013) was never pursued.

Centurion beach armoured recovery vehicle: FV4018

The Sherman beach armoured recovery vehicle (BARV or, occasionally, beach ARV) had shown itself to be enormously useful during the amphibious landing stages of the Normandy campaign following D-Day. With the turret and gun removed and the hull sides raised to allow the vehicle to wade in up to 96in (8ft) of water, the BARV was used for recovering drowned or disabled tanks.

The Sherman BARVs remained in service until the late 1950s and when the time came to replace them it seemed logical to use the hull of the then-current main battle tank, the Centurion. The Fording Trials Branch of REME produced a mock-up along the lines of the Sherman, using a surplus Centurion tug. The hull was extended by about 5ft and a large rope-cushioned pusher pad was installed at the front; this pad was replaced by a hardwood nosing block to reduce the danger of damaging landing craft. A prototype was demonstrated at the Amphibious Trials and Training Unit (ATTU) at Instow in 1959 and by the end of 1960 a batch of twelve had been constructed at the Royal Ordnance Factory Leeds using Mk 3 hulls. The overall height, with the armoured hull extension, was 140in and the vehicle was capable of wading in 114in (9ft 6in) of water; all-up weight was 40 tons.

When the Army's amphibious capability was phased out in favour of the Royal Marines, the BARVs were similarly reassigned, and two are known to have been taken to the Falkland Islands in 1982 along with the Task Force.

Conqueror armoured recovery vehicle: FV219, FV222

The Conqueror presented a whole new set of problems and, with a combat weight approaching 65 tons, it was obvious that the Centurion ARV would not be adequate as a recovery vehicle. Clearly, a Conqueror-based vehicle would need to be developed.

An early proposal for a heavy armoured recovery vehicle (FV209) based on the universal tank was abandoned, but the hull was used for trialling the winches that would eventually be used in the Conqueror ARV Mk 1 (FV219), with the winch capable of a massive 45-ton pull. In March 1953 it was suggested that three Conqueror ARV prototypes be constructed by Vickers-Armstrong for troop trials. A mock-up was inspected in June 1953. The layout of the vehicle was similar to the Centurion ARV, with a fixed superstructure in place of the turret to house the crew and the winch, and with the driver placed in the crew compartment, as in the Conqueror gun tank. A small crane jib could be attached to the rear spade. The first prototype, weighing some 57 tons, began trials two years later, with the second passed to REME for trials during 1957. Although it had originally been planned that twenty vehicles would be constructed, with the first due for delivery at the beginning of 1955 and the last by late 1959, it seems that just eight were completed before attention turned to the improved Mk 2 variant (FV222).

The two versions can easily be distinguished by the revised shape of the hull, and particularly the use of a long sloping glacis plate. Mechanically the Mk 2 was similar to the Mk 1, but there were improvements to the winching arrangements, and the driver was moved to the crew compartment. A mock-up was shown to potential users in October 1955, with the design approved for production immediately. By August 1957 it was said that all the production drawings were complete, albeit production did not actually start until September 1959. The first vehicle entered service in 1960 and a total of twenty were constructed.

Curiously, the Conqueror was far more successful as an ARV than it had been as a gun tank and a number saw longer service than the Conquerors themselves. Until it was superseded by the Challenger armoured repair and recovery vehicle (CR ARRV), the FV222 was the most powerful recovery vehicle used by the British Army.

Chieftain armoured recovery vehicle: FV4204

Developed jointly by the Military Engineering Experimental Establishment (MEXE) and the Royal Ordnance Factory Leeds, the FV4204 Chieftain armoured recovery vehicle was intended to replace the earlier Centurion ARV, which was beginning to show its age. The final version of the requirement document was issued in late 1964, with the first two prototypes delivered to REME at Bordon in 1971. The results of the trials necessitated that the hydraulic system be completely redesigned, and the vehicle finally went into production at Vickers Defence Systems in 1974, with the first example accepted for service in 1976. It was eventually superseded by the Challenger armoured repair and recovery vehicle (CR ARRV), development of which started in 1985.

The Chieftain ARV was constructed on the hull of the Chieftain Mk 5 and utilised the same power-pack, suspension and auxiliary generator. The main winch, which was an electro-hydraulic double-capstan design, rated for a straight pull of 30 tons, was placed at the front of the vehicle, alongside the driver; there was also a 3-ton auxiliary winch. Power for both winches was provided by the main engine via a power take-off. At the front of the vehicle there was a 'dozer blade, operated on a pair of hydraulic arms, which could also be used as an anchor when winching.

The later Chieftain ARRV, which was also supplied to Iran, was fitted with an Atlas hydraulic crane on the rear deck.

Challenger armoured repair and recovery vehicle

Although the Challenger 1 main battle tank was constructed by the Royal Ordnance Factory Leeds, a competition for the related armoured repair and recovery vehicle (ARRV) in 1985 saw the work awarded to Vickers Defence Systems at Newcastle-upon-Tyne. The first contract, awarded in that same year, called for thirty vehicles to be constructed, on a fixed price contract. Subsequent contracts increased the number of vehicles to eighty, of which thirty-two were built at Newcastle and forty-eight at Leeds, the Royal Ordnance Factory having been taken over by Vickers Defence Systems in 1986.

Weighing 67.5 tonnes and with a spare Challenger power-pack carried on the rear deck, the Challenger ARRV (sometimes known as the CR ARRV) uses the same hull and automotive systems as the gun tank and is equipped with a Rotzler Treibmatic hydraulic double-capstan winch capable of a straight pull of 62 tons. Other equipment includes an Atlas hydraulic crane, installed on the rear deck and capable of lifting a complete Challenger power-pack, a stabiliser blade for the crane and a hydraulic 'dozer blade which doubles as an earth anchor during winching operations. A special trailer to be towed behind the ARRV has been designed to carry a Challenger, Chieftain or Warrior power-pack.

Despite Challenger 1 having been superseded by the much improved Challenger 2 MBT, at the time of writing (spring 2011), the Challenger ARRV remains the British Army's standard heavy armoured recovery vehicle for the foreseeable future.

Samson armoured recovery vehicle: FV106

Designated FV106, Samson is the armoured recovery vehicle variant of the Alvis CVR(T) family and is capable of recovering light armoured and logistical vehicles on difficult terrain. The vehicle shares automotive components with others in the series and was developed jointly by Alvis and MEXE. The first prototype appeared in the early 1970s, with production starting in 1977.

Weighing a shade under 20 tons, Samson is constructed around an aluminium-armoured hull similar to that used for the Spartan armoured personnel carrier (APC), but with a 20-ton hydraulic winch installed in the rear compartment, driven via a power take-off from the main engine. Two spade-type earth anchors at the rear of the hull can be released manually to assist in vehicle placement during winching operations. A small jib crane can be erected on the rear of the hull to enable Samson to be used for light lifting duties, and there are stowage facilities inside the hull for a range of recovery tools and attachments.

Maximum speed on the road is 45mph, with 4mph available during wading operations.

Warrior repair and recovery vehicle: FV512, FV513

The Warrior armoured repair and recovery vehicle is constructed around the basic hull and superstructure of the Warrior infantry combat vehicle and has been produced in two variants: the combat repair vehicle (MCRV, FV512), which is equipped with a crane but no winch, and the recovery and repair vehicle (MRVR, FV513), which has both crane and winch, the latter fitted inside the hull at the rear. A one-man rotating turret is fitted, mounting a co-axial 7.62mm Hughes chain gun. Combat weight is a shade under 30 tons, and the vehicle is capable of 45mph on the road.

Inside the hull there is sufficient space for a five-man crew – commander, driver, gunner and two fitters – and a fully equipped workshop. The hydraulic crane, which is carried on the left-hand side of the superstructure, is rated at 6,500kg, and has sufficient capacity to lift and replace the power-pack of either a Challenger main battle tank or a Warrior infantry combat vehicle. The winch is capable of a 20,000kg single pull, and there is a large hydraulic earth anchor at the rear.

The British Army has taken delivery of a total of 110 examples of the MCRV and sixty-seven MRVRs; they are operated by REME detachments in armoured infantry battalions.

Alvis Hippo BRV

Originally described as the 'future beach recovery vehicle' (FBRV), and designed very much in the same style as the Centurion and Sherman BARVs of the Second World War, the Hippo armoured beach recovery vehicle (known now simply as a BRV) was introduced in 2003. The development of the Hippo was undertaken by Hagglünds, a wholly owned subsidiary of Alvis, using the chassis and running gear of the Krauss-Maffei Leopard 1A5. Like its predecessors, the Hippo is designed to rescue stranded vehicles, both armoured and soft skin, from landing beaches, under fire if conditions so dictate; it is also able to push a beached LCU 10 landing craft back into the sea, even when loaded.

With the Leopard's gun and turret removed, a raised superstructure has been attached to the top of the hull, allowing the vehicle to operate in up to 116in (9ft 8in) of water. Rubber nose blocks prevent damage to either vehicle during pushing operations, and the Hippo can also tow vehicles weighing up to 50 tons. The original 12-cylinder MTU MB 873 turbocharged diesel engine of the Leopard has been retained, but for this application the gearing of the transmission was lowered to increase tractive force. Combined with a maximum weight of 50 tons, this has reduced the vehicle's top speed to 20mph.

Four Hippos are in service with the Royal Marines, one each on the two Albion Class amphibious assault vessels HMS Albion and HMS Bulwark, and two based at the Royal Marines' Testing and Training Centre at Instow.

Battle-damaged Centurions were sometimes converted to tugs in Korea, but the official Centurion ARV did not appear until 1951. A total of 170 vehicles were constructed, with the first entering service in 1956. (*IWM, MVE 43138/1*)

This privately owned Centurion Mk 2 ARV shows the low, purposeful appearance of the machine. The majority of the vehicles produced were conversions of the Mk 5 hull; conversion work was carried out by Vickers-Armstrong and ROF Woolwich. (*Simon Thomson*)

Like the Sherman beach armoured recovery vehicles (BARV) that had been used so successfully on the Normandy beaches, the Centurion BARV was designed to be able to recover vehicles that had become bogged down during landing operations. The hull was extended to allow operation in water up to 96in (8ft) deep. (*Warehouse Collection*)

A Centurion BARV coming to the rescue of a drowned amphibious ¾-ton Land Rover (FV18051). The first production BARV was trialled at Instow in 1960, with a total of twelve vehicles constructed using Centurion Mk 3 hulls. (*Warehouse Collection*)

While the Conqueror was not much of a success as a gun tank, the power of the winch meant that the armoured recovery teams of the Royal Electrical & Mechanical Engineers (REME) considered it among the best recovery vehicles they had seen. The Mk 2 (seen here) was improved in several respects. (*Warehouse Collection*)

Semi-overhead view of the Conqueror ARV Mk 2. The winch is situated in the fixed superstructure that replaced the turret, and the components of a small jib can be seen on the rear deck, designed to be fitted to the rear spade. (*Warehouse Collection*)

A Chieftain Mk 5 ARRV (FV4204) using its rear-mounted hydraulic crane to remove the engine cover of a Challenger 1 tank; the earlier ARV variant lacked the crane. (*Warehouse Collection*)

In 1985 development work started on an armoured repair and recovery (ARRV) variant of the Challenger 1 to replace the Chieftain ARV. A total of eighty examples were constructed and the Challenger ARRV (or CR ARRV) remains the British Army's current vehicle of this type. (*Simon Thomson*)

The Challenger ARRV is capable of towing vehicles weighing up to 68 tons at a speed of 20mph. The vehicle mounts a hydraulic crane and can tow a trailer on which can be carried a spare Challenger or Warrior power-pack. The Challenger ARRV is the first vehicle of its type specifically designed to enable a large amount of repair work to be carried out in the field. (*Vickers Defence Systems*)

Challenger ARRV showing the front-mounted 'dozer blade, which can also act as an earth anchor during winching operations, and as a stabiliser for the Atlas crane mounted on the rear deck. There are two winches, both of which can be operated from the driver's position while the hull is closed down. (*Simon Thomson*)

Samson (FV106) is the armoured recovery vehicle variant of the CVR(T) series. A 20-ton hydraulic winch is installed in the rear compartment, and twin spade anchors are carried at the rear. A jib crane can be erected on the hull. (*Alvis Vehicles*)

Photographed during trials, this Samson ARV (registration 03SP38), which is almost certainly the first example produced, is using its winch to recover a disabled FV432 armoured personnel carrier. (*Tank Museum*)

A Warrior ARRV showing the purpose-designed trailer carrying a spare armoured vehicle power-pack. (*GKN Defence*)

Based on the German *Leopard* 1A5 tank, the Hagglünds Hippo is described as a beach recovery vehicle and replaces the earlier Centurion BARV. Four of these vehicles are in service with the Royal Marines. (*Alvis Vehicles*)

The Hippo is capable of operating in water to a depth of 116in (9ft 8in) and has a top speed of 20mph. (*Alvis Vehicles*)

ARMOURED VEHICLE LAUNCHED BRIDGES

The first armoured bridging vehicles appeared at the end of the First World War using modified Mk V** heavy tanks. By 1925 a 16ft-span light girder bridge had been designed that could be pushed into place by the Vickers medium tank. Subsequently, Matilda infantry tanks were used in the same way, pushing longer-span bridge sections across gaps of up to 80ft. However, the modern concept of a vehicle-launched bridge did not appear until the mid-1930s, when Covenanter, Valentine and, later, Churchill tanks were fitted with a mechanically deployed scissors bridge for the bridgelayer role. During the post-war years this method of bridging became well established, with both Centurion and Chieftain chassis used as the basis of bridgelayers.

Current nomenclature describes these machines as 'armoured vehicle launched bridges'.

Centurion bridgelayer: FV4002

The design work that led to the Centurion bridgelayer started in 1946, although it had originally been intended to use the chassis of the FV200 series universal tank. Early experiments involved mounting a lattice steel framework on to the hull of Centurion prototype number three to test the manoeuvrability of what was inevitably a somewhat extended vehicle. The cancellation of the FV200 project meant that the work took a back seat but in 1952 a mock-up bridge was mounted on a Centurion Mk 1 hull and by 1956 a working prototype had been constructed by Hudswell Clark of Leeds using a Mk 7 chassis.

User trials of this, and a second, modified, prototype, were completed by September 1958, when it was decided that the production vehicles would be based on redundant Mk 5 hulls, reworked to bring them up to Mk 7 standard. The gun and turret were removed and a roof plate was fitted over the turret ring; the resulting vehicle weighed 49.6 tons with the bridge in place and was capable of a top speed of 20mph on the road. The first pre-production Centurion bridgelayer was completed by the Royal Ordnance Factory Leeds in early 1960 and, following acceptance trials, production proper started in 1961 and continued until 1963. A number of export orders followed.

The hydraulic bridge-launching equipment was powered by a Rolls-Royce B40 four-cylinder auxiliary engine placed in the fighting compartment alongside the radio operator; the auxiliary engine was connected by a propeller shaft to a Towler Brothers hydraulic pump. The Class 80 'bridge, tank, number 6' consisted of four identical aluminium-alloy quarter trackways joined together in pairs to give a 52ft-long bridge (maximum span 42ft); a lifting jib was provided on the launch arm to assist in the assembly process. Lifting brackets on the trackways were used to attach the bridge to the launch arm, and for travelling the bridge was carried along the length of the hull in an inverted position. During the launch operation, which took two minutes, the bridge was simply lifted from its stowed position, rotated through 180 degrees and placed on the ground behind the tank, at which point the launch vehicle was disengaged. In-fill panels were carried on the sides of the hull, designed to be placed across the gap between the two longitudinal bridge components to allow smaller, wheeled vehicles to cross. Recovery was more-or-less the reverse of the launch process, taking four minutes.

Although the Centurion bridgelayer was more than up to the task assigned to it, its sheer size and lack of manoeuvrability meant that it was unable to keep up with the gun tanks in urban areas, particularly in West Germany. However, the vehicle remained in service until 1974, when it was replaced by the more versatile Chieftain bridgelayer.

Centurion ARK: FV4016

Some Churchill tanks had been modified to act as armoured ramp carriers (ARK) – effectively, a rapid assault bridge – during the Second World War and these remained in service into the immediate post-war years, although by the late 1950s they were showing their age. A replacement was designed by the Fighting Vehicle Research & Development Establishment (FVRDE) using the hull of a Centurion Mk 5 from which the turret and gun were removed, with a roof plate used to cover the turret ring. The commander was relocated inside the hull alongside the driver. Ramps were attached to either end of the hull, and a trackway was fitted to the top of the hull; in operation, the tank was driven into the centre of a ditch or trench, or up against a sea wall, and the ramps were folded out hydraulically at either end to form a Class 80 continuous bridge, with the tank remaining in place during use. In travelling configuration, the length of the ARK exceeded 34ft; when deployed, the 81ft-long bridge gave a useful span of 75ft and a width of 14ft across the ramps, which was wide enough to accommodate the massive width of the Conqueror.

A variant of the ARK, described as the Centurion ARK mobile pier (CAMP), was developed to provide a central pier support in the centre of a waterway for two 'number 6' tank bridges. The ramps were omitted, leaving only the central trackway in place.

Constructed by the Royal Ordnance Factory Leeds, the Centurion ARKs remained in service until 1975, by which time this type of equipment was considered obsolete.

Chieftain armoured bridge launcher: FV4205

Designed as a replacement for both the Centurion bridgelayer and the Centurion ARK, work on the development of the Centurion armoured vehicle launched bridge (AVLB) started in 1962 at the Military Engineering Experimental Establishment (MEXE), with assistance from Lockheed Precision Products and Tubes (Birmingham)

Hydraulic Controls. The prototype was constructed by the Royal Ordnance Factory Leeds, but considerable redesign work was required before production could commence. This meant that the first examples did not enter service until 1974.

Like the Centurion, the Chieftain AVLB was based on the hull of a gun tank, with the turret and gun removed but instead of the Centurion's 'number 6' tank bridge, the Chieftain was equipped to carry either a 'number 8' Class 60 tank bridge, with a clear span of 75ft, or a 'number 9' Class 80 boom-launch bridge, with a single span of 40ft; both bridges were constructed from aluminium-zinc-magnesium alloy and nickel-alloy 'maraging' steel (so-called maraging or martenistic ageing steels possess superior strength and toughness without losing malleability). The 'number 9' bridge was carried and launched in much the same way as the 'number 6' bridge on the Centurion, while the 'number 8' bridge was carried in a folded position and launched scissor-style across a gap from the front of the vehicle; a bank-sensing device was fitted to the leading ramp. Combat weight carrying a 'number 8' bridge was 52.35 tons, and the vehicle was capable of a maximum speed of 30mph on the road.

Unlike the Centurion, the hydraulic equipment was driven via a power take-off on the main engine, with the launching operation taking between three and five minutes. After deploying the bridge the launch vehicle was disengaged; recovery, which could be effected from either end, took about ten minutes.

In operation, it was common practice to carry the 'number 9' bridge on the Chieftain hull, and for this to be accompanied by a Scammell Constructor tractor and semi-trailer which was used to carry the 'number 8' bridge.

Titan armoured vehicle launched bridge

Based on the Challenger 2 hull and power-pack, Titan is the British Army's current armoured bridging vehicle and it has completely replaced the Chieftain armoured bridge launcher. Titan was developed by Vickers Defence Systems in 2001, with the first prototypes appearing two years later.

Titan can operate across a wider range of terrain conditions than the previous Chieftain AVLB and, for example, can launch a British Army 'number 10' 85ft-long scissors-style steel bridge in two minutes, and one or a pair of 39ft-long 'number 12' bridges in ninety seconds. It can also lay multiple combination bridges. The bridge, or bridges, are carried across the hull and are launched from the front of the vehicle, allowing the launcher to be disengaged once deployment is complete. All launch and recovery operations can be completed from under armour. A track-width 'dozer blade or mine plough can also be fitted to the front of the hull.

The vehicle entered service with the Royal Engineers in 2006. It has been described as the 'world's fastest bridgelayer', forming the armoured component of what is known as the British Army's modular bridging system (BR90). The last of thirty-three examples constructed was delivered in 2008.

Photographed at the Royal Engineers' Museum in Chatham, Kent, this FV4002 Centurion bridgelayer carries a 'number 6' tank bridge. The launching operation, which involves rotating the bridge through 180 degrees, took just two minutes. (*Warehouse Collection*)

A Centurion bridgelayer with the bridge removed to show the launch arm and the rear support on which the bridge is carried. (*Warehouse Collection*)

The Centurion ARK (FV4016) was designed to remain in place to permit tanks and other vehicles to cross a ditch or other shallow obstacle. The ramp sections were folded out to form a Class 80 bridge with a 75ft span. (*IWM, MH 9803*)

A Chieftain armoured bridge launcher with the 'number 8' scissors bridge. The bridge launching operation, which took a maximum of five minutes, is shown at the halfway stage. (*Warehouse Collection*)

The Chieftain bridgelayer could also carry the 'number 9' bridge, capable of spanning a 40ft gap by rotating the bridge through 180 degrees using hydraulic rams. (*Tank Museum*)

The BAE Systems Titan is the British Army's current armoured bridgelayer, although these days it is described as an 'armoured vehicle launched bridge'. Based on the Challenger 2 hull and power-pack, Titan is seen here carrying a pair of 39ft-long 'number 12' bridges, which can be launched in ninety seconds. (*Simon Thomson*)

Titan can also be used to carry the 'number 10' bridge, a scissors design capable of spanning an 85ft gap, with a launch time of around two minutes. It can also lay multiple combination bridges. (*BAE Systems*)

ARMOURED ENGINEERS' VEHICLES

The first armoured engineers' vehicles were developed during the Second World War after the unsuccessful assault at Dieppe in August 1942. Designated 'assault vehicle, Royal Engineers' (AVRE) and based on the Churchill hull, the vehicle was designed to provide a sufficient degree of protection to allow working under fire. AVREs were used to equip the 1st Assault Brigade of the 79th Armoured Division in time for use on the D-Day beaches.

The Second World War AVRE was a very adaptable vehicle, armed with a 290mm *pétard* spigot mortar in the turret, capable of firing a 40lb 'dustbin' charge some 230 yards. Brackets were fitted on the front and sides of the hull to accept all kinds of attachments, including the small box-girder bridge (SBG), various hessian and canvas mat-laying devices designed to prevent tanks from becoming bogged down in soft ground, demolition-charge placing devices and the fascine bundle, which was used as an aid to crossing ditches and trenches. AVREs were also used to push 'skid' Bailey bridges into position.

Many of the original Churchill AVREs, which were based on the Churchill Mk III and Mk IV chassis, remained in service during the post-war years, but were eventually superseded by an upgraded version based on the Churchill Mk VII. New vehicles were also developed using the Centurion and Chieftain chassis.

Churchill AVRE: FV3903

Described as 'Churchill VII AVRE, tank, infantry, dozer' and allocated the identifier FV3903, this particular variant of the Churchill AVRE was a post-war conversion of the heavily armoured Churchill Mk VII, featuring two major modifications. The standard gun of the Churchill Mk VII was replaced by a breech-loading 165mm howitzer, which was a considerable improvement on the 290mm muzzle-loaded *pétard* mortar used on earlier Churchill-based AVREs, and a full-width 'dozer blade was installed across the front of the hull, with the support arms pivoted at around the mid-point.

This version of the Churchill AVRE started to enter service in 1954 and served with the British Army well into the mid-1960s, eventually being replaced by the Centurion AVRE.

Centurion AVRE: FV4003

The first proposals for a Centurion-based AVRE date back to September 1950, with a prototype produced at the Fighting Vehicles Proving Establishment (FVPE) using the hull of Centurion prototype number four (02BA58), together with the turret of a Centurion Mk 1 into which had been installed a 95mm howitzer. A full-width hydraulically operated 'dozer blade, produced by the Newcastle company T.B. Pearson & Sons, was fitted to the nose.

Problems with development and a shortage of materials meant that the definitive prototype, using the hull of the Centurion Mk 7 gun tank, was not produced until August 1957. Production did not get under way until 1963, by this time using the Mk 5 hull fitted with a 165mm L9A1 gun firing a 64lb demolition charge. Just forty were constructed, some of which were based on the hulls of Centurion Mk 12 artillery observation posts fitted with a Pearson's mine plough in place of the standard 'dozer blade, but retaining the 105mm gun.

A jettisonable 15-ton four-wheel trailer (FV2721A) was also developed for use with the Centurion AVRE to carry either a fascine bundle or trackway.

Centurion 'dozer: FV4019

Although the initial requirement dated back to 1958, it was a further three years or so before the Centurion 'dozer (FV4019) started to enter service, replacing the ageing Centaur 'dozer which, of course, lacked a turret. The Mk 5 gun tank was selected for the conversion, being fitted with a hydraulically operated Pearson's dozer blade, identical to that fitted to the Centurion AVRE. The hydraulic pump, manufactured by H.M. Hobson, was driven by the main engine. The main gun remained operative, although there was some reduction in the number of rounds that could be stowed, and both Mk 5 tanks, equipped with the 17-pounder (76.2mm) gun, and Mk 5/1 tanks, the latter with the 105mm gun, were converted. With the 'dozer equipment in place, the weight was increased to 52.6 tons and the Centurion became somewhat nose heavy; the additional weight precluded up-armouring and made the hull unstable when firing to the rear.

Centurion 'dozers were also exported to Australia and Denmark.

Chieftain AVRE

The original Chieftain AVRE project (FV4203) was abandoned due to lack of funding, and it wasn't until 1986 that a number of early Chieftain gun tanks were converted to the AVRE role for use in Germany. The work was carried out by the Engineer Workshops of 40 Army Engineer Support Group at Willich, with a total of twelve vehicles converted. Following a period of trials, the converted Chieftains started to enter service in 1987, replacing the older Centurion AVREs.

The conversion entailed removing the turret and main gun and installing an armoured roof over the turret ring. Trackways were fitted along the length of the hull to carry up to five rolls of Class 60 trackway or three fascine bundles, by this time consisting of plastic pipe sections. The 'dozer blade was the standard item produced by A.P. Precision Hydraulics of Liverpool for use with the Chieftain gun tank; the standard blade could also be replaced by a Pearson's mine-plough blade. These interim vehicles were subsequently superseded by an improved version developed by Vickers Defence Systems from 1989.

Two prototypes of the Vickers-designed Chieftain AVRE were trialled at Bovington in 1991, with production of forty-eight examples commencing the following year and continuing until 1994. As with the earlier interim design, the turret and gun were removed but in this instance they were replaced by an armoured-steel superstructure. A hydraulic winch was mounted on the rear of the hull, together with a pair of so-called 'hampers', designed to carry and launch the fascines or trackways, or to carry other engineers' equipment; a hydraulic loader was provided to lift the fascines or trackways into position. At the front of the hull was a removable Pearson mine-plough or standard Pearson 'dozer blade; when not mounted on the vehicle, the 'dozer or plough could be stowed on the rear 'hamper'.

Chieftain 'dozer

A.P. Precision Hydraulics also produced a bulldozer kit that could be attached to any 'mark' of Chieftain to provide a basic tactical earth-moving vehicle, on which the main gun remained operative. The kit consisted of an electrically operated hydraulic power-pack, joystick controls, aluminium 'dozer blade and interconnecting wiring and hydraulic pipework. The 'dozer assembly was mounted to the front towing eyes.

Although the kit was adopted by the British Army, and in theory could easily be moved from one vehicle to another, it was said to take up to six hours to fit and was never entirely satisfactory.

Combat engineer tractor (CET): FV180

Designed by the Military Engineering Experimental Establishment (MEXE) and constructed by the Royal Ordnance Factory Nottingham, the combat engineer tractor (CET) was developed in response to General Staff Target 26 (GST26), which described an engineers' vehicle that combined the mobility of a tracked vehicle with armoured protection, the performance of a heavy-duty earth-mover and the ability to swim. Although definitely not a 'tank' under any standard definition of the term, the CET was designed to provide armoured engineer support to a battle group, and was able to perform many of the roles previously assigned to earlier armoured engineers' vehicles that were based on tank chassis.

The origins of the vehicle can be traced back to 1963, when Caterpillar, GKN and Vickers were each asked for proposals to meet GST26. None of the proposals was felt to be acceptable and in 1965 MEXE came up with its own design, with a pair of prototypes subsequently constructed by the Royal Ordnance Factory Nottingham. At the time there was some discussion regarding joint development of the vehicle between Britain, West Germany (as was) and France, but by 1970 both France and West Germany had decided to go their own way. Following a series of trials, the vehicle was redesigned to improve the amphibious capability, and costs were contained by using as many standard commercial components as possible. A further seven prototypes were constructed and delivered during 1973/74 and, following further modifications, the vehicle was finally accepted for production in July 1975, although it was to be almost another three years before the first example rolled out of the Royal Ordnance Factory.

Constructed around an aluminium welded hull and weighing somewhere in the order of 17.7 tons, the CET is normally driven with the bucket to the front, which places the two crew members on the left-hand side; both crew positions are reversible and duplicate driving and bucket controls are provided. The rear engine and transmission are mounted on the right, with the final-drive components at the front. The engine is a Rolls-Royce C6TFR six-cylinder turbocharged diesel, driving the front sprockets and the rubber-faced steel tracks, through a David Brown TN26 four-speed manual gearbox, arranged to provide all four gears in either direction. In normal driving mode steering is by means of a controlled differential but for 'dozing the driver has an independent brake-and-clutch system. Amphibious operation is effected by means of a pair of Dowty Hydrojet units, driven from an engine power take-off, and twin trim-boards, complete with Hycafloat buoyancy units (from FPT Industries, Portsmouth), are provided at the rear to aid swimming capability; polyurethane foam blocks are also carried in the bucket to further aid buoyancy. Suspension of the four road wheels is effected by torsion bars; a fifth road wheel acts as a track idler.

The bucket has a capacity of 1.88 cubic yards and is carried on a hydraulic arm in such a way as to allow it to be used for digging and 'dozing; a fixed jib crane can be erected in the bucket, using the winch for lifting purposes. The 'dozer blade is also designed to act as an earth anchor for the two-speed 8-ton winch installed inside the hull. A second earth anchor is carried on the roof of the hull, designed, if necessary, to be rocket-deployed, and to assist the vehicle to exit riverbanks, etc. A nuclear-biological-chemical (NBC) filter pack is installed at the front of the vehicle to allow operation under hostile battlefield conditions. A pusher bar can be fitted to allow the CET to be used to launch a bridge, and the vehicle can also be used to tow the Giant Viper mine clearance system.

The last of 143 CETs for the British Army was produced in March 1981 but a further fifteen were delivered to India between 1988 and 1990, and fifty-four went to Singapore during 1993–95.

Trojan obstacle-crossing vehicle

Announced in 2001 and entering service from 2007, the BAE Systems' Trojan is a multi-functional armoured engineers' vehicle based on the hull and running gear of the Challenger 2 gun tank. While under development it was designated as the 'future engineer tank' by the Ministry of Defence, and it is described by its makers as 'a unique complex obstacle-crossing vehicle'.

A wrist-action hydraulic excavator, fitted with a large, 1.3 cubic yard bucket with a lifting capacity of 16.5 tons, is pivoted on the front of the hull on the right-hand side, and is normally carried to the rear; the bucket can be replaced with an impact hammer or earth auger. A hydraulic power take-off is provided, operating at $2,030lbf/in^2$ with a flow rate of 7gal/min. Across the front of the vehicle is an interchangeable full-width 'dozer blade or mine-plough, while rails on the rear deck allow the carrying of a midi-fascine bundle (26.25ft).

Weighing more than 62 tons, Trojan can be used for mine clearance, excavating and 'dozing, ditch crossing and lifting; it can also be used to carry engineers' stores and equipment. All of these operations can be carried out with the three-man crew remaining under full armoured protection, and a 7.62mm machine gun is fitted to the hull for self-defence. Mine-clearance work can be carried out using the mine-plough, or by firing Python explosive mine-clearing hose from a towed trailer; the cleared path can be automatically marked from under armour using an on-board obstacle marking system. A fascine bundle can be launched into small gaps to allow ditches or other obstructions to be crossed, and the vehicle can dig using the excavator and/or the 'dozer.

A total of thirty-three vehicles are currently in service, at a cost of £4.217 million each, and Trojan was first deployed operationally in Afghanistan in 2009.

Constructed around the hull of the Churchill Mk VII, which was a veteran of the Second World War, the FV3903 Churchill AVRE (Armoured Vehicle Royal Engineers) started to enter service in 1954 and served alongside the Centurion AVRE. The muzzle-loading 290mm mortar was capable of firing a demolition charge, and a full-width 'dozer blade could be fitted to the front of the hull. (*Warehouse Collection*)

The Centurion AVRE was based on the Mk 5 hull and mounted a 165mm gun firing a 64lb demolition charge. Just forty were constructed. This example has been fitted with additional spaced armour. (*Simon Thomson*)

A close-up of the Centurion AVRE 'dozer blade. This vehicle was photographed during the Gulf War, by which time it was probably more than thirty-five years old. (*Warehouse Collection*)

A Chieftain AVRE, complete with plastic-pipe fascine in the front carriers; note the hydraulic loading equipment at the rear and the mine-plough at the front. This photograph was taken at the British Army Training Unit at Suffield in Canada (BATUS). (*R. Ridley*)

While no one would describe the combat engineer tractor (CET) as a tank, it has earned its place here by virtue of carrying out operations previously executed by so-called engineers' tanks. The large front-mounted bucket can be used for both digging and 'dozing, and a crane jib can be erected in the bucket for lifting operations. (*Warehouse Collection*)

Although the CET is normally driven with the bucket towards the front, duplicate controls allow it to be driven and operated from either end. (*Simon Thomson*)

The CET is capable of amphibious operation using a pair of Dowty Hydrojet units, driven from an engine power take-off. Twin trim-boards, with Hycafloat buoyancy units, are provided at the rear, and polyurethane foam blocks are also carried in the bucket to further aid buoyancy. (*Warehouse Collection*)

The obstacle-crossing Trojan is a multi-functional engineers' vehicle, able to carry out mine clearance, ditch crossing, excavating, 'dozing and lifting operations. It is seen here with the 1.3 cubic yard hydraulic bucket extended to the front. (*BAE Systems*)

Delivery of the Trojan started in 2007, and the British Army currently has thirty-three vehicles, purchased at a total cost of more than £139 million. (*BAE Systems*)

The Trojan is capable of a maximum speed of 35mph and can cope with a variety of terrain. (*BAE Systems*)

Chapter Five

British-made Tanks
not used by the British Army

The first tanks were designed by William Foster of Lincoln and Walter Wilson during the First World War, with production of the machine entrusted to Fosters and the Metropolitan Amalgamated Railway Carriage & Wagon Company of Wednesbury, a company that was taken over by Vickers-Armstrong in 1919. Until 1936 Vickers was the only company producing tanks in Britain.

During the Second World War, when it was very much a case of 'all hands to the pump', companies such as Vauxhall, Nuffield and Leyland all became involved in tank production, but in the post-war years British tank production has been in the hands of two companies: Vickers-Armstrong (later Vickers Defence Systems) and the government-owned Royal Ordnance Factories (ROF). When Vickers took over the tank-production facilities of the Royal Ordnance Factories in 1986, it became Britain's only tank manufacturer, and when Vickers was itself taken over by Alvis in 2002, the new company dominated the production of armoured fighting vehicles in this country and, as BAE Systems, went on to take over General Dynamics in the USA, manufacturers of the US Army's Abrams main battle tank.

Although Vickers played a significant role in the development and production of the Centurion and the Conqueror, and was subsequently wholly responsible for the Challenger 2, much of the design and production work on the Chieftain and the Challenger 1 went to the Royal Ordnance Factories. In order to survive in the business of building tanks, Vickers secured contracts to modify and manufacture British tanks for the export market. For example, along with the Royal Ordnance Factory Leeds, Vickers was involved in the production of Chieftains for Iran and Kuwait, and also assisted in the further development of the Chieftain 800 and 900, which had originally been designed at Leeds, as the possible basis for selling surplus Chieftains to Pakistan.

However, in the late 1950s Vickers decided that it would once again produce tanks to its own designs for export to developing countries, a business that the company had found very profitable during the interwar years.

CHIEFTAIN 800 and 900

Dating from around 1981, Chieftain 800 and Chieftain 900 were produced as a private venture by the Royal Ordnance Factory Leeds. The intention was to upgrade Chieftain by incorporating improvements to firepower, protection and mobility, while keeping the weight below 56 tons. Two development vehicles were constructed using a late model Chieftain chassis. The original Leyland L60 engine, which had been heavily criticised for its lack of performance and reliability, was ousted in favour of a twelve-cylinder Rolls-Royce Condor coupled to a David Brown TN12/1000 six-speed transmission. The engine of the Chieftain 800 was rated at 800bhp, while that of the Chieftain 900 was rated at 900bhp.

Major improvements to the hull included the addition of Chobham composite armour, and there was also a new welded-steel turret mounting the ROF 120mm L11A5 rifled tank gun.

There were no subsequent sales but both vehicles have survived, at least in part. The Chieftain 900 is at the Tank Museum, while the turret from the Chieftain 800 was recently put up for auction by Witham Specialist Vehicles, having been mounted on a standard Chieftain hull by Marconi Avionics as a trials vehicle in 1993.

SHIR 1 and 2

Under the former Shah's regime, Iran had purchased large numbers of Chieftain gun tanks and armoured recovery vehicles. In late 1974 the Royal Ordnance Factory Leeds received a huge contract for two variants of an upgraded Chieftain dubbed Shir (the name was derived from 'SHah' of 'IRan'). With a total of 1,350 tanks covered by the contract, this was the largest single export order received by ROF during peacetime.

Shir 1 (FV4030/2), of which there were to be 125 examples, was a late production Chieftain with modified suspension that doubled the amount of wheel travel available, a Rolls-Royce CV12 TCA Condor diesel engine, coupled to a David Brown TN37 automatic transmission, and the addition of a Marconi Avionics computer sighting system. In early 1979, following the revolution in Iran, the contract was cancelled while production was under way but the Shir 1 was subsequently modified to become the Khalid main battle tank.

Designed by the Royal Armament Research and Development Establishment (RARDE), Shir 2 (FV4030/3) was a far more ambitious project. Although the power-pack was identical to that fitted to Shir 1, the hull was completely new and, for the first time on a British armoured fighting vehicle (AFV), incorporated hydro-pneumatic suspension. There was also a new angular turret which, like the hull, incorporated Chobham composite armour. By the time the contract was cancelled, RARDE had started trials on at least two prototype vehicles.

Shir 2 eventually evolved into the Challenger 1. Ironically, the British Army's obsolete Challengers have ended up in Jordan, not quite Iran's next-door neighbour but not far away.

KHALID

In November 1979 the Hashemite Kingdom of Jordan placed a £266 million order with the Royal Ordnance Factory Leeds for 278 Khalid main battle tanks, with delivery scheduled to begin in 1981. Based on the hull of the Chieftain Mk 5, Khalid was essentially a further development of the FV4030/2 that had been produced for Iran, but with minor modifications specified by the Jordan Arab Army. Modifications to the rear of the hull allowed the Leyland L60 engine to be replaced by a Rolls-Royce CV12 TCA Condor unit (later to become the Perkins Condor), connected to a David Brown TN37 four-speed automatic transmission driving the rear sprockets – in effect, giving the Chieftain hull the running gear of the later Challenger 1. The suspension was a development of that used on the Chieftain but with increased travel.

The main armament was the ROF 120mm L11A5 rifled tank gun, installed in conjunction with a co-axial 7.62mm machine gun; a second 7.62mm machine gun was mounted on the commander's cupola. A new fire-control system was fitted.

The changes resulted in a combat weight of 55 tons and the tank had a top speed on the road of 32mph.

ROF RO2004 LIGHT TANK

Designed and constructed by the Royal Ordnance Factory Leeds at the end of the 1980s, and predating the company's 1986 take-over by Vickers Defence Systems, the RO2004 was the light tank of the company's RO2000 series general-purpose tracked chassis. The vehicle was equipped with a 105mm L7 rifled gun with a bustle-mounted autoloader, and there was also a co-axial Hughes 7.62mm chain gun. The hull was designed for a three-man crew, and was protected by what the company described as a 'dynamic armour system', with the addition of appliqué armour on the sides and face of the turret. It was claimed that this would provide protection against both chemical-energy and kinetic-energy rounds to a standard not generally available from vehicles in this market sector.

The engine was a Perkins TV8-640, eight-cylinder turbocharged diesel, producing 320bhp from 10,488cc, and driving the front sprockets through a Self Changing Gears T320 automatic six-speed transmission, a combination able to drive the 21.5-ton vehicle to a maximum speed of 35mph. Independent transverse torsion-bar suspension was provided for each of the five road wheels and ROF claimed that a hydro-pneumatic suspension system was in development.

Other variants on the same chassis included an armoured mortar vehicle, armoured personnel carrier and 105mm or 122mm self-propelled gun. The series was intended to be sold to customers in the Middle East and to developing nations elsewhere, but, although prototypes were said to have been delivered to Egypt for competitive trials and were tested in excess of 6,000 miles 'over some of the most arduous conditions in the world', there is no evidence that any sales were actually achieved.

VICKERS MAIN BATTLE TANKS

In the late 1950s Vickers had designed a light tank armed with a 20-pounder (84mm) gun as a private venture. This never progressed beyond the drawing board but in 1961 Vickers signed an agreement with the Indian government under which the company would help to establish a tank-manufacturing facility near Madras to construct the Vijayanta main battle tank – the name means 'victorious' or 'conqueror'. The agreement also provided that Vickers would produce the first ninety examples before passing production across to India in 1965.

This was the first of a series of main battle tanks that culminated in the appearance of the Vickers/FMC VFM5, developed jointly by Vickers Defence Systems and the American FMC Corporation in 1986.

Vickers MBT Mk 1

The prototype of what was effectively the Vickers main battle tank (VMBT) Mk 1 was completed in 1963 and the tank entered service in 1965. Designed to be as simple and cost-effective as possible, the vehicle was constructed around an all-steel welded hull with a maximum armour thickness of 80mm, and with accommodation for a driver, commander, gunner and loader. The combat weight of the vehicle was 37.5 tons. Armaments included the 105mm L7 weapon of the Centurion, together with a co-axial 7.62mm machine gun, a second similar gun mounted on the turret, plus a turret-mounted 0.5in ranging machine gun.

Production vehicles were powered by the same type of Leyland L60 multi-fuel engine as the Chieftain, driving through a David Brown TN12 six-speed semi-automatic epicyclic transmission. With a combat weight some 15 per cent lighter than the Chieftain, the L60 gave a top speed on the road in the order of 30mph, combined with a cross-country performance that was said to be at least equal to any contemporary main battle tank. The VMBT Mk 1 was also offered with a GM Detroit-Diesel 12V-71T two-stroke diesel engine, and other engines were also considered, including a pair of Rolls-Royce K60 units, a German MAN diesel and a US-built Teledyne Continental engine.

By the time production ended in India in 1983, some 2,200 vehicles had been constructed in Britain and India, with seventy delivered to Kuwait between 1970 and 1972.

Vickers MBT Mk 2

Dating from 1968, the VMBT Mk 2 was an improved version of the original machine, with a new turret front, redesigned frontal aspect to the hull, upgraded tracks and more powerful engine. There was also provision for fitting a pair of launchers for the BAC Swingfire wire-guided anti-tank missile on either side of the turret rear. There was no series production.

Vickers MBT Mk 3

Incorporating many of the improvements seen in the Mk 2, including, for the first time, a commander's cupola, the prototype for the VMBT Mk 3 was completed in 1975. The original Leyland L60 engine was replaced by a GM Detroit-Diesel 12V-71T turbocharged two-stroke diesel giving a slight improvement to the maximum speed on the road, to 32mph. The twelve-cylinder Rolls-Royce CV12 TCE Condor engine was also available as an option. No changes were made to the weapons systems, but there were improvements to the sighting and fire-control equipment.

Production got under way three years later, with a number of vehicles delivered to Kenya and Nigeria. There was also an armoured repair and recovery vehicle (ARRV) version, supplied to Tanzania and Nigeria in small numbers, as well as an armoured bridgelayer, also purchased by Nigeria.

Production continued until the end of the 1980s, by which time the tank had been upgraded to VMBT Mk 3(I) – meaning 'improved' – by the addition of a new transmission system, muzzle-reference system and changes to the hull.

Vickers Valiant MBT Mk 4

With a name borrowed from one of the least successful tank designs of the Second World War – or perhaps, more likely, one which nodded in recognition of Britain's four-engined bomber, part of the Royal Air Force's Cold War V-bomber nuclear force of the 1950–60s – the VMBT Mk 4 was designated Valiant. The design work started in 1977, with the first prototype shown at the British Army Equipment Exhibition (BAEE) in 1980.

Constructed around an aluminium hull and an angular, welded steel universal turret, the Valiant was designed to exploit the development of Chobham composite armour, which enabled the weight to be kept down to around 47 tons without compromising protection. The first prototype was armed with the familiar ROF 105mm L7A3 rifled tank gun, but this eventually gave way to a 120mm L11 weapon;

the Rheinmetall 120mm smooth-bore gun, as fitted to the German *Leopard* main battle tank, was also available as an option. The secondary armament consisted of a co-axial Hughes 7.62mm chain gun, together with either a 12.7mm or 0.5in machine gun mounted on the commander's cupola. At the time the target-acquisition and fire-control systems were considered to be among the best in the world.

Power was provided by a Rolls-Royce CV12 TCA Condor diesel, developing 1,000bhp and driving through a six-speed David Brown TN12/1000 fully automatic transmission.

It is difficult to determine whether any were actually sold, but the turret went on to be used on the VMBT Mk 7.

Vickers MBT Mk 5 or Vickers/FMC VFM5

With its distinctively rounded shape, the Vickers MBT Mk 5 – also known as the Vickers/FMC VFM5 – was a lightweight main battle tank intended for export to developing countries. It was the result of a collaboration between Vickers Defence Systems and the American FMC Corporation, following the signing of an agreement in 1985.

Constructed around a completely new hull of aluminium armour and weighing just 20 tons, the VFM5 was designed to be air-portable by C-130 Hercules or C-141 Starlifter transport aircraft, and yet the advanced armour was said to provide higher levels of protection than that available in other tanks of a comparable weight or class. The turret was designed to accept either the ROF 105mm rifled tank gun, the Rheinmetall 105mm super low-recoil gun or the US 105mm M68, firing a wide range of ammunition types. A 7.62mm machine gun was mounted co-axially with the main weapon, together with a 0.5in or 7.62mm machine gun at the loader's station. Computerised fire-control equipment was supplied by Marconi Avionics.

A GM Detroit-Diesel 6V-92TA turbocharged diesel engine was installed in the rear, accessible via a fold-down hatch, and coupled to a Lockheed-Martin HPMT-500-3 fully automatic transmission, driving the rear sprockets. Suspension was provided by virtue of trailing arms and torsion bars.

The first prototype appeared at the British Army Equipment Exhibition (BAEE) in 1986, but it seems that the design failed to attract any sales.

Vickers MBT Mk 7

The VMBT Mk 7, or Mk 7/2, was a joint collaboration between Vickers Defence Systems and the German company Krauss-Maffei, and was a logical development of the Mk 4, attempting to combine the firepower and turret system of the Mk 4 Valiant with the improved mobility of the Mk 5. Work started in 1984, with the first prototype appearing a year later.

It was essentially the chassis of the Krauss-Maffei *Leopard 2*, on which was fitted an updated version of the turret of the Valiant, mounting an ROF 120mm L11 rifled gun designed to fire fin-stabilised anti-tank rounds; Rheinmetall 120mm and Giat 120mm guns, both of smooth-bore configuration, were also offered as alternatives. Secondary armament included a pair of 7.62mm machine guns, one installed co-axially with the main gun. The VMBT Mk 7 was one of the first tanks to be fitted with computer-based digital fire- and gun-control systems, the equipment being supplied by Marconi Avionics.

The engine was the German MTU MB 873, a twelve-cylinder turbocharged diesel with a power output of 1,500bhp, coupled to a Renk HSWL four-speed automatic transmission driving the rear sprockets. In common with the Leopard 2, the hull was constructed from spaced, multi-layered composite armour, and rode on torsion-bar suspension. The turret was of welded steel, with a layer of appliqué Chobham armour on the front and sides. All-up combat weight was 54 tons and the maximum road speed was 45mph, with a range of 340 miles.

Despite Vickers' claim that the VMBT Mk 7 could out-perform any existing tank in its class, just one vehicle was constructed and it made its first public appearance at the British Army Equipment Exhibition in 1988. There were no sales, but the VMBT Mk 7 ultimately led to the development of the British Army's Challenger 2.

VICKERS/NORINCO NVH-1 MCV

Developed and marketed jointly by Vickers Defence Systems and China North Industries (NORINCO), the NVH-1 mechanised combat vehicle (MCV) consisted of the welded-steel hull of the Chinese H1 armoured personnel carrier (APC), on which was fitted a two-man Vickers power-operated turret, mounting a 30mm RARDEN cannon. A 25mm Hughes chain gun was mounted co-axially with the main weapon. There was sufficient space inside the hull to accommodate nine fully equipped men, including the driver and commander.

The first prototype appeared in 1986, powered by a Deutz BF8L air-cooled turbocharged diesel engine, coupled to a four-speed manual transmission. There was no series production.

VICKERS MCV

The Vickers mechanised combat vehicle (MCV) may have been designed as an entry in the contest for the British Ministry of Defence (MoD) MCV-80 project that ultimately resulted in the production of the GKN Warrior.

Intended to combine a Vickers power-operated turret with the proven hull and superstructure of the American Bradley M2 infantry vehicle, it was powered by a Cummins VTA-903T turbocharged diesel, coupled to a General Electric HMPT-500 fully automatic transmission. Vickers proposed two variants; the first was an armoured personnel carrier armed with a 30mm RARDEN cannon or Bushmaster M242 chain gun in a small turret, while the second was a light tank, mounting the ROF 105mm L7 low-recoil gun, together with a 7.62mm Hughes chain gun in a larger angular turret. Both provided accommodation for ten men in the hull.

Illustrations in the sales literature produced at the time of the vehicle's launch show what are clearly scale models, so it may well be that no full-sized vehicles were produced.

VICKERS ARMOURED REPAIR AND RECOVERY VEHICLE

In response to a request for three such vehicles from Kenya in 1977, the Vickers main battle tank Mk 3 was modified to provide an armoured repair and recovery vehicle (ARRV), either with or without a hydraulic crane for changing AFV power-packs in the field.

With the turret removed, a 25-ton capstan winch was installed in the left-hand side of the hull, alongside the driver, with a large spade anchor fitted at the front. A cradle could be fitted to the rear deck to enable the vehicle to carry an AFV power-pack, and buyers could also ask for an auxiliary winch to be installed. A 7.62mm general purpose machine gun (GPMG) was mounted on the roof. There was a choice of engine: either the Rolls-Royce V12 1200A or a General Motors 12V-71T, in both cases coupled to a David Brown TN12 automatic transmission.

The first examples were delivered in 1981, with seven vehicles ultimately delivered to Kenya, ten to Nigeria and a number to Tanzania.

VICKERS ARMOURED BRIDGELAYER

During the late 1970s Vickers started designing an armoured bridgelayer known as the VAB, using the hull and superstructure of the Vickers Mk 3 main battle tank. The vehicle carried a Class 60/70 Biber horizontally launched bridge with an overall length of 44ft; the bridge was carried in an inverted position across the hull, and launched hydraulically before being detached from the launch vehicle. After use the bridge could be recovered and rotated back on to the vehicle hull.

Although the VAB was ostensibly designed for possible sale to the Middle East, the only buyer has been Nigeria, which took a total of twelve of these vehicles.

First appearing in 1981, Chieftain 900 was a private venture by ROF Leeds. Based on an upgraded Chieftain fitted with a Rolls-Royce engine, it included Chobham composite armour and was fitted with a 120mm rifled tank gun. Despite these various improvements to what was already a formidable tank, there was no series production. (*Warehouse Collection*)

(*Above and right*) The Khalid main battle tank (FV4211) was essentially an improvement on the Iranian Shir 1, orders for which had been cancelled when the Shah was deposed in 1979. Based on the hull of the Chieftain Mk 5, but with the Leyland L60 engine replaced by a Rolls-Royce, which necessitated a longer engine compartment, it was constructed by ROF Leeds, with a total of 278 vehicles supplied to Jordan from 1981. (*Warehouse Collection*)

The hull of the ROF Leeds-designed RO2004 light tank. Built largely for export to the Middle East and to developing nations, the RO2004 family included an armoured mortar vehicle, an armoured personnel carrier, and 105mm and 122mm self-propelled guns, all constructed on a common chassis. (*Warehouse Collection*)

The Indian Army Vijayanta main battle tank was also offered for broader sale under the designation Vickers main battle tank Mk 1 (VMBT Mk 1). (*Vickers Defence Systems*)

There was no series production of the VMBT Mk 2, but the Mk 3 was much improved. The main gun was not changed, but the Leyland L60 engine was replaced by a GM Detroit-Diesel 12V-71T turbocharged two-stroke diesel, which brought improved performance. (*Vickers Defence Systems*)

The VMBT Mk 3(I) was further improved by changes to the hull and turret, and by the addition of a new transmission and muzzle-reference system. (*Vickers Defence Systems*)

The VMBT Mk 5 – or Vickers/FMC VFM5 – was a joint project between Vickers Defence Systems and the American FMC Corporation dating from 1985. Sadly no actual sales were made. (*Vickers Defence Systems*)

The rounded shape of the lightweight aluminium hull of the VMBT Mk 5 made for a very impressive machine and, with a turret designed to accept either the ROF 105mm rifled tank gun, the Rheinmetall 105mm super low-recoil gun or the US 105mm M68, it was capable of an equally impressive performance. (*Vickers Defence Systems*)

With work starting in 1984, the VMBT Mk 7 was a joint collaboration between Vickers Defence Systems and Krauss-Maffei, and was intended to combine the versatile firepower and turret system of the Mk 4 Valiant with the improved mobility of the Mk 5. Despite being shown at BAEE in 1988, there were no sales, but the design eventually evolved into the Challenger 2. (*Vickers Defence Systems*)

A joint venture between Vickers Defence Systems and China North Industries (NORINCO), the NVH-1 mechanised combat vehicle was based on the welded-steel hull of the Chinese H1 armoured personnel carrier (APC), to which was fitted a two-man Vickers power-operated turret with a 30mm RARDEN cannon. (*Vickers Defence Systems*)

Scale model of the Vickers mechanised combat vehicle, which may have been intended as a contender for the British Army's MCV-80 project, eventually won by the GKN Warrior. (*Vickers Defence Systems*)

Based on the chassis of the VMBT Mk 3, the Vickers armoured repair and recovery vehicle was supplied to Kenya, Nigeria and Tanzania. The turret was removed and a 25-ton winch fitted into the hull; the vehicle could be specified with or without a rear-mounted hydraulic crane. (*Vickers Defence Systems*)